Language, Ethics and Animal Life

Language, Ethics and Animal Life

Wittgenstein and Beyond

Edited by
Niklas Forsberg, Mikel Burley and Nora Hämäläinen

B L O O M S B U R Y
NEW YORK • LONDON • NEW DELHI • SYDNEY

Bloomsbury Academic
An imprint of Bloomsbury Publishing Inc

1385 Broadway   50 Bedford Square
New York   London
NY 10018   WC1B 3DP
USA   UK

www.bloomsbury.com

Bloomsbury is a registered trade mark of Bloomsbury Publishing Plc

First published 2012
First published in paperback 2014

Library of Congress Cataloging-in-Publication Data
Language, ethics, and animal life : Wittgenstein and beyond / edited by
Niklas Forsberg, Mikel Burley, and Nora H?m?l?inen. – 1 [edition].
pages cm
Includes bibliographical references and index.
ISBN 978-1-62892-236-3 (pbk. : alk. paper) – ISBN 978-1-4411-4055-5 (hardback : alk.
paper) 1. Wittgenstein, Ludwig, 1889-1951–Congresses. 2. Animals–Philosophy–
Congresses. 3. Language and languages–Philosophy–Congresses.
I. Forsberg, Niklas, editor.
B3376.W564L36 2014
179'.3–dc23
2014017412

ISBN: HB: 978-1-4411-4055-5
PB: 978-1-6289-2236-3
ePub: 978-1-4411-5568-9
ePDF: 978-1-4411-6462-9

Typeset by Newgen Knowledge Works (P) Ltd., Chennai, India

Contents

Acknowledgements

Forerunners of most of the essays that constitute this volume were first presented at the inaugural conference of the Nordic Wittgenstein Society in Uppsala, Sweden, in March 2010. We are grateful to all the speakers and other delegates who participated in that very successful conference and to the Bank of Sweden Tercentenary Foundation for financial support.

Two anonymous reviewers for Bloomsbury provided helpful comments which enabled us to refine the volume as a whole and prompted us to commission a further contribution from Nancy Baker, to whom we are also much indebted.

Tremendous thanks are due to Ika Österblad for supplying the marvellous cover illustration and design, and Haaris Naqvi at Bloomsbury for seeing this volume through to publication.

All the essays are published here for the first time, with two exceptions. Alice Crary's paper, 'W. G. Sebald and the Ethics of Narrative', previously appeared in *Constellations: An International Journal of Critical and Democratic Theory*, 19(3) (2012). We and Alice Crary are grateful to Wiley-Blackwell and the journal's editors for permission to republish the essay. Stefano Di Brisco's essay, 'Second Nature and Animal Life', is a substantially revised and longer version of his article by the same title that first appeared in *Between the Species: An Online Journal for the Study of Philosophy and Animals*, 13(10) (2010), 118–31, which is a journal sponsored by the Philosophy Department and Digital Commons at California Polytechnic State University <http://digitalcommons.calpoly.edu/bts/>. We and Stefano Di Brisco are grateful to these sponsors and to the journal's editor-in-chief, Joseph Lynch, for permission to re-use material from the previously published version of the paper.

Niklas Forsberg
Mikel Burley
Nora Hämäläinen
February 2012

Notes on Contributors

Nancy E. Baker is Professor of Philosophy at Sarah Lawrence College where she has taught Wittgenstein's *Philosophical Investigations* for 37 years. She has contributed a paper to *Feminist Interpretations of Wittgenstein,* and has written on Wittgenstein's *On Certainty* and 'The Role of "Confusion" in Wittgenstein's Later Work'. She is currently finishing a non-academic book to be titled, *Why Can't A Cat Retrieve? Wittgenstein, the Siamese Cat and I.*

Mikel Burley is Lecturer in Religion and Philosophy at the University of Leeds. He is the author of *Contemplating Religious Forms of Life: Wittgenstein and D. Z. Phillips* (Continuum, 2012) and two books on Indian philosophy. His recent articles include 'Winch and Wittgenstein on Moral Harm and Absolute Safety' (*International Journal for Philosophy of Religion,* 2010), 'Emotion and Anecdote in Philosophical Argument: The Case of Havi Carel's *Illness*' (*Metaphilosophy,* 2011) and 'Believing in Reincarnation' (*Philosophy,* 2012).

David Cockburn is Professor of Philosophy at University of Wales: Trinity Saint David. His publications include *Other Human Beings* (Macmillan, 1990), *Other Times: Philosophical Perspectives on Past, Present and Future* (Cambridge University Press, 1997), *An Introduction to the Philosophy of Mind* (Palgrave, 2001) and a wide range of papers on themes in philosophy of mind, ethics, Wittgenstein, philosophy of religion and philosophy of time.

Alice Crary is Associate Professor in Philosophy at the New School for Social Research in New York. Her main research and teaching interests are moral philosophy, the philosophy of Wittgenstein, philosophy and literature, feminist theory, and ethics and animals. She is the author of *Beyond Moral Judgment* (Harvard University Press, 2007), the editor of *Wittgenstein and the Moral Life: Essays in Honor of Cora Diamond* (MIT Press, 2007) and the co-editor of *Reading Cavell* (Routledge, 2006) and *The New Wittgenstein* (Routledge, 2000). She is currently completing a book on animals and ethics entitled *Inside Ethics.*

Stefano Di Brisco obtained his PhD from Sapienza – University of Rome in 2011 with a thesis on moral philosophy. He has been visiting scholar at the New School for Social Research in New York. His interests include Wittgenstein and moral philosophy, ethics, and philosophy and literature.

Niklas Forsberg is Lecturer in the Department of Philosophy at Uppsala University, Sweden. He is currently completing a monograph entitled *Language Lost and Found* which treats the thought that some philosophical problems arise due to our having lost the sense of our own language and how that problem is dealt with in recent discussions of the relationship between philosophy and literature. In this monograph, Iris Murdoch is his main conversation partner. He has previously written on Wittgenstein, Cavell, Murdoch, Austin and Derrida.

Rami Gudovitch wrote his dissertation at Columbia University, entitled *A Nominalist Conception of the Extent of the Intentional*. His main fields of specialization include metaphysics, early analytic philosophy and the philosophy of mind. His current research projects involve the development of a position that denies the supervenience of the mental on the physical, and investigating the role of authority in education. He is presently completing his post-doc at Haifa University and teaching philosophy at the Interdisciplinary Center in Hertzliya and at Ben-Gurion University.

Ylva Gustafsson is a PhD student in the Department of Philosophy at Åbo Akademi University, Finland. She is currently working on her doctoral thesis where she discusses questions on interpersonal understanding.

Nora Hämäläinen is a post-doctoral researcher and temporary lecturer in philosophy at the University of Helsinki, Finland. In 2009–11 she was editor-in-chief of the Helsinki-based cultural weekly *Ny Tid*, which presents a mixture of literature, arts, politics and culture. She has written about ethics, literature and the nature of moral theory.

Julia Hermann received her PhD from the Department of Political and Social Sciences of the European University Institute and is now Lecturer in Social Philosophy at Maastricht University. In autumn 2008, she was a Visiting Student Researcher at the Philosophy Department of the University of California, Berkeley. In the academic year 2010/11, she worked as a Teaching Fellow in the *European Master's Programme in Human Rights and Democratisation* in Venice. Her main research interests include the philosophy of the later Wittgenstein, moral justification, moral knowledge, moral education and human rights.

Camilla Kronqvist defended her doctoral dissertation in philosophy in 2008 at Åbo Akademi University in Finland. Since then she has held different research and teaching positions at the same university. In her dissertation she investigated emotional, interpersonal and moral aspects of the concept of love. Articles on the same theme appear in *Inquiry* and *Philosophical Investigations*. She is the co-editor, together with Ylva Gustafsson and Michael McEachrane of *Emotions and Understanding: Wittgensteinian Perspectives* (Basingstoke: Palgrave, 2009), and continues working on questions relating to emotions in her post-doctoral research.

Olli Lagerspetz has been Senior Lecturer in Philosophy at Åbo Akademi University since 2005. He was the Head of Philosophy from 2007 to 2010. In addition to Åbo, he has studied at the University of Illinois at Urbana-Champaign, and was Lecturer in Philosophy at the University of Wales, Swansea, from 1992 to 1998. He has published a book on the concept of trust and a book (in Swedish) on the clean and the dirty, as well as articles on ethics, philosophical anthropology and the philosophy of mind.

Anne Le Goff is currently completing her PhD at the University of Picardie in France. She works on the philosophy of second nature of John McDowell, and is interested in non-reductionist naturalist approaches to human life as well as animal life. She is the co-editor of a philosophical reader of *Mind and World* by McDowell in French, and the translator of many philosophical texts from German and English, including *Jewish Philosophy as a Guide to Life* by Hilary Putnam.

Pär Segerdahl is Associate Professor of Philosophy at Uppsala University in Sweden. His background is in philosophy of language. He has published several inquiries into linguistic theory, for example, in *Language Use* (Macmillan, 1996). He often uses examples from ape language research, above all in *Kanzi's Primal Language* (Palgrave, 2005), written with ape language researchers William Fields and Sue Savage-Rumbaugh. Pär Segerdahl currently works at the Centre for Research Ethics and Bioethics (CRB) in Uppsala.

Different Forms of Forms of Life:
A Philosophical Introduction

Niklas Forsberg

1. Three examples

First example: A short sequence from the movie *Alien: Resurrection*.[1] We see a man lying with his chest under a vehicle repairing it. He is whistling. He lies in a mechanical wheelchair which has a backrest that can be tilted backwards. Above him, another man is standing on a ledge, playing with a knife – sounding like a monkey. With all the marks of an intentional action the man on the ledge 'drops' the knife so that it falls and penetrates the leg of the man under the vehicle, but there is no immediate reaction from him. Instead, a woman named Call says: 'What is wrong with you!' 'Just a little target-practicing,' the man on the ledge replies, 'Vriees isn't complaining.' Vriees slides out, looks at his leg and says: 'Goddammit! Johner, you son of a bitch!' 'Ah, come on man,' Johner replies, 'you didn't feel a thing.' Johner makes a phony monkey laugh. 'You are an inbred motherfucker, you know that?' Call says. She looks at him with anger in her eyes, pulls the knife out of Vriees's leg and breaks it – an act that offends Johner deeply.

Second example: A friend of mine grew up in a house outside the city with farms neighbouring their house, so they were always near animals. Farm keeping, hunting and fishing were everyday activities for them and the transformation from animal to food came, as it were, naturally to them. He and his sister had two pet rabbits, 'Snow White' and 'Sooty', which were very dear to them. But, as most pet animals grow old long before we humans do, the time came when my friend's family regarded the rabbits as needing to be done away with. My friend's father took care of that and, shockingly, prepared a meal out of them. At the dinner table, the children cheerfully fought over the cooked rabbit parts: 'I want Snow White!'; 'I want Sooty's leg,' they said and ate their pets with smiles on their faces.

Third example: Georgina (Figure 1.1).

She's dead. We killed her, me and my former girlfriend 'J'. Georgina had a heart condition. That's why we had her put to death. For nearly two years, she was medicated with diuretic medicine that helped remove fluid from her lungs and a heart medicine that supposedly widened her artery so that her weakening heart could move her blood

---

[1] *Alien: Resurrection*, directed by Jean-Pierre Jeunet (Twentieth Century Fox, 1997).

Figure 1.1 Photograph by Niklas Forsberg.

more easily. But eventually the muscle would no longer do it. We knew that. As her heart weakened, more fluid assembled in her lungs and she started to cough. During her last days she had trouble breathing, so she stretched her neck up – as if looking towards the sky, or at J – thus enabling some air to get into, if not fill, her lungs.

Finally the day came. J called me on the phone and said that Georgina wasn't doing so well and asked if I wanted to come with them to the veterinarian. When I arrived at their apartment, it became evident that Georgina was in a very poor condition. She was slow, had no appetite and her breathing was laboured. She barely had the lungs to cough with and once she even lost her balance. It was very likely that we would take Georgina to the veterinarian but not back.

The examination went fairly quickly. Georgina was in very bad shape and we were confronted with something of a dilemma. We could place Georgina in intensive care for a week or so – a treatment which may give her a couple more months – or we could put her to sleep right now. To take her home again was not a possibility. One choice. Two options. Neither of them good. Neither one acceptable.

The decision concerning Georgina's existence was J's to make. (In our community dogs are owned and she was hers.) The veterinarian was extremely helpful and she let us take all the time we needed. The examination room had a door facing a parking lot so we went outside with Georgina to think things through. In an attempt not to upset her we made a deal not to cry in front of her, taking turns to go aside and dispose of tears. As we were trying to make up our minds about Georgina's existence she walked slowly back and forth between us. She clearly knew something was wrong. She knew that *we* were not feeling well. The dog we were about to put to sleep was comforting us.

J asked the vet if she could see the room Georgina would be placed in if we decided to give intensive care a chance. When J came back outside, she shook her head. The matter was settled. Georgina would die of sorrow and fear if not from her heart condition in that cage. Perhaps her history as an abandoned stray dog formed her, but she did not enjoy solitude. She hated it.

So the decision was made and Georgina's passing was about to begin. The vet assured us that this was a good thing to do. 'This is the right decision. You are doing it for her sake.' Sometimes, sincere and comforting words ring false. Of course, it requires a peculiar perspective on things to describe the injection of poison as 'the right thing' or 'for her sake'. One of Georgina's forelegs was shaved. A needle was inserted and taped into place. This gave her some very understandable stress, so the vet said that we should let her walk around a bit so that she could calm down – which she did, on three legs with lowered head. She was placed in J's lap and injected. The bladder relaxed and urine spilled over J's jeans as an emblem of the order of things. Georgina's final transformation, from tense to completely soft, had taken place and J and I went to our separate homes – J with an empty leash in her hands.

2. Philosophical drama and argumentation

The three examples above are not unproblematic from a philosophical perspective. First of all, they are dramatic, emotionally charged and it is not evident from the start what philosophical work they are meant to do. I have placed the three examples in a particular order. The first is perhaps easiest to incorporate into contemporary academic philosophy. There seems to be one single act performed there (the brute 'dropping' his knife) which we may or may not condemn, pass judgement upon for this or that reason; say, it was mean because it is always mean to bully. We can also imagine a philosopher (of a particular bent) arguing that it was not really an act to be condemned for moral reasons because the man in the wheelchair had no feeling in his legs, and so (in an 'objective' sense) no harm was done because harm always has a physical root (in, say, pain). Strange as it may seem, there are philosophers who wilfully and proudly state that they succumb to repugnant conclusions – thus attempting to imply that their own hearts and egos are the subordinates of 'rationality' and 'objectivity'.

The second example is more complex from that perspective. Surely, we can single out one action to pass judgement upon in that too – eating one's pets – but the nature of this 'action' is, at least seemingly, more difficult to explain. In the case of the stew made of bunnies, the fact that it *is* real (not fictional) may appear to put a particular form of pressure upon us. For example, it becomes harder to condemn the action without also saying something about the persons involved. Simply put: the action's surroundings cannot be circumvented without considerable distortion of the nature of the action. It matters that they were kids, that they grew up close to animals, that they were used to seeing animals slaughtered and prepared for meals, that this particular meal was cooked by their father (a trustworthy and upright man) and so on. But it also matters that when my friend, many years later, tells the story he finds it at once absurd, comical, tragic and bizarre – something very remote from the life he leads today. It thus brings into view how complicated a reference to a 'form of life' is – a phrase that is recurring in nearly all of the essays in this volume.[2] If we want to evaluate this action morally, the response 'Eating animals he knew was simply a part of his previous form of life, and hence it was correct by that "standard"; and then his form of life shifted, and now it is no longer so' handsomely seems to beg the question. We might use such talk to describe that a shift has occurred, and perhaps also to describe what kind of shift we have in front of us. But does it give us a sufficient foothold to make moral judgements? Does it *explain* anything about this particular region of our moral lives in language? The philosophical reference to the fact that two forms of life differ, that they can and do change, has to do a kind of work other than explaining what is right and wrong.

The third example is also the longest. I take this example to be even harder to turn into something that most contemporary academic moral philosophers might accept as 'philosophy'. One might say that the first example is fictional and that the second is real but anecdotal (which already makes it problematic). But the third is personal, emotional and near to confessional. Thus it clearly violates some common assumptions of what philosophy is (at least of the contemporary kind, which looks towards science for an ideal of intellectual rigour): it is not objective (enough?), and it may also be considered too emotional, too personal and indeed too detailed to give us *any* kind of generality. Surprisingly perhaps, the fictional example thus seems to lend itself more easily to academic discussions than the real. One might easily picture someone saying that in that case it is at least clear that we are talking about an example, and an example is always an example *of* something else, something general.

Of course, someone might argue that my third example includes a typical philosophical element: a moral dilemma. Should we put the dog to sleep now or attempt to postpone death by means of intensive care? And then we might quarrel and give reasons for and against our different views – working under the assumption that, for near to logical reasons, one of these actions *must* be at least a lesser evil than the other and that that makes it 'right' (if not good). But can the details of that tale be so removed without significant loss? Is the sense of necessity in the starting point that one of the actions has to be a lesser evil illusory? For us who were involved, it was evident that we were not really confronted with a case of good vs. bad, or right vs. wrong. It was bad vs. bad, wrong vs. wrong, intolerable vs. intolerable. We might have reached some kind of mental peace in assuring ourselves that 'it was the only thing we could do', but that doesn't make it right and it most certainly doesn't make it 'good'. The only good outcome in this case was a healthy dog. But the good was not an option. So we act upon conscience, but not without remorse.

It is true that we did what we thought best and that, in some sense, we were 'calculating' – trying to sense the weight of different alternatives. But that does not nullify the fact that for me and J, the description 'the right action' was not exactly

---

[2] For Wittgenstein's use of the phrase, see for example, his *Philosophical Investigations*, trans. G. E. M. Anscombe, 2nd edn (Oxford: Blackwell, 1958), §§19, 23, 242 and pp. 174 and 226.

false, but void or hollow, misplaced, carrying no genuine weight. That we did what we thought was best does not remove the weight on our shoulders from an event that was bad through and through.

If it is argued (or silently assumed) that there has to be one sense of the event according to which we can rest assured that we did the right thing (in an objective sense), does not that include a sort of denial of the sense of tragedy that I and J experienced? A philosophy of an abstract, theoretical, intellectualized kind – one that builds on what Alice Crary, in 'W. G. Sebald and the Ethics of Narrative', calls a narrow conception of what an argument is and of what 'rational' is supposed to mean – might claim that our sense of tragedy was *merely* a sense, not really real, and as such 'nothing more' than an emotional reaction; something philosophy must (try to) disregard. But what if emotions are not mere outbursts of the unreflective ego – outbursts, that is, in a nearly 'biological' sense – but are, rather, part of our understanding? I would like to suggest that the tormenting feelings of regret and remorse, of uncertainty and sorrow, *are* part of an adequate understanding of this situation. If you remove them and turn the death of this animal into an example or an intellectual puzzle, you have not reached the core of the moral concerns involved but distorted our reality to such an extent that it is only in a diluted sense that we are talking about 'the same event'.

The ways in which the habitual distinction between emotions and understanding is problematic are further highlighted and discussed in Ylva Gustafsson's 'What is Altruism?' and Camilla Kronqvist's 'Talking about Emotion'. Gustafsson shows us that we misunderstand the nature of altruism if we do not attend to how that concept is intertwined with more complex aspects of our moral lives in language, such as what it means to 'have a life', and that we often go wrong in our theoretical accounts of altruism (and related concepts) because it is very easy to think of the concept of understanding in purely epistemological terms. Similarly, Kronqvist shows us that we have a lot to gain from thinking about what it is that we are *doing* (performing certain actions, not merely stating things) when we are talking about emotion in relation to humans and animals. The multifarious ways in which talking about emotion comes into our lives, displays clearly how one can go wrong if one adopts a wholly naturalistic and scientistic perspective on the question about the nature of emotions. There is a form of scientism that underlies a surprisingly common idea: namely, that any true account of morality and of humans' relations to animals must be based on descriptions in which our emotional involvement with one another has been deliberately left out, or set aside as irrelevant and even dangerous. Such accounts, however, run the risk of achieving only partial accuracy. At worst, they give us a distortion. Kronqvist and Gustafsson can help us see that the idea that emotions need to be disregarded goes along with, on the one hand, the attempt to disable the importance of context and, on the other, the desire to turn real and oftentimes messy human concerns into something manageable in a 'purely intellectual' and 'rational' way. In saying this, I do not mean to imply that morality can be reduced to emotions, as if our emotions are the foundation on which our moral judgements rest. Rather, it is an attempt to rediscover a broadened understanding of 'understanding'.

The death of Georgina shows us, at least, that human weakness is forced upon us by the facts of reality. One bad, regrettable, decision *had* to be made. I feel inclined to say that the death of this animal can show that the desire to intellectualize, to turn a death into a dilemma, to calculate pros and cons, is completely understandable – called for, if you wish. But perhaps we turn to these resources of our intellect not so much to find out what really is the case, but to make the unbearable facts of life bearable. Is it possible that we sometimes turn to arguments to hide our own weaknesses? Do we at times turn to abstract theorizing in order not to face reality? And, if so, how does that affect philosophy, how *we* do philosophy? (These are questions, not answers. I'm asking, not asserting.) This much seems to be true: philosophical clarity may require a form of writing that enables us to absorb the intimate details of our lives in language.

Raimond Gaita has argued that the use of records of personal engagements with animals is problematic *if* they are meant to serve as counterarguments to scientific or philosophical theories that adhere to a narrow view of what counts as objective and what counts as argumentation.[3] Personal anecdotes and other narratives may be useful if we are to achieve a more comprehensive view of the human–animal relation, but they cannot constitute a rebuttal of science on science's terms. In Gaita's view, their primary role must be understood in terms of conceptual elucidation. In 'Honour, Dignity and the Realm of Meaning', Nora Hämäläinen dwells on this thought, expands it and shows how the idea of conceptual investigations needs to be open to conceptual changes and to the fact that language is constantly developing.

The claim that philosophy needs richer, fuller and more detailed and nuanced depictions of our lives with animals – which we see in Crary's discussion of Sebald as well as in Hämäläinen's discussion of Gaita, Anscombe, Coetzee and Lidman – should not merely be seen as more attentive or emotionally sensitive alternatives to other more abstract and theoretical philosophical theories. Rather, the turn to literature includes a reconsideration of the whole field under investigation. It is an invitation to reconsider *what* the field is. Crary and Hämäläinen are not exactly contesting the validity of the arguments of alternative theories. Rather, literature comes in as a reminder of *what* our moral world is. It helps to bring into view how complex and multifarious the world we share with animals is, and that many moral theories go wrong, not because their arguments are unsound, but because they rely on a reductive picture of what our world is like and of what counts as rational and what count as arguments. The turn to literature is a call for a more adequate description and understanding of our world and our language, a more austere form of realism, than most philosophical theories about these matters accommodate. Put otherwise, W. G. Sebald's 'Austerlitz' and J. M. Coetzee's 'David Lurie' and 'Elizabeth Costello' are far more real than most 'agents' we encounter in analytic philosophy. 'Peter', 'Mary', 'Smith' and 'John' – doubtless the most commonly employed names in this region of philosophical exemplification – are variables, not persons.

Anecdotes of various sorts may also seem to provide a much too particular outlook on things. ('*You* claim *this* happened, *once*! So what?') And it is true that for scientific purposes the personal and the particular remain problematic. It is obvious that the

---

[3]　For a further discussion of these themes, see Cora Diamond, 'Anything but Argument?' in *The Realistic Spirit: Wittgenstein, Philosophy and the Mind* (Cambridge, MA: MIT Press, 1991), 291–308.

repeatability of the experiment is of vital importance in scientific research. But one should not therefore conclude that generality always must come together with facts of quantity and repeatability. As Pär Segerdahl's 'Humanizing Nonhumans' shows, philosophical clarity requires something other than the type of certainty, probability, corroboration and what not, that the repeatability of an experiment may provide. Is it, for example, really sensible to ask for evidence for the other's ability to speak when he is talking to you? When he is rebuking you, calling you a monster? Scientific tests generally involve an idea of the expected outcome – you need to know *what* you are testing – so the scientist, more often than not, knows what he or she is looking for when testing or experimenting. This is not to say that all scientific research runs smoothly or that science never encounters the unexpected, as if all science could be practiced mechanically. I merely mean that experiments and tests are moulded, designed, for a purpose. The question is whether this is true about philosophy too. As Segerdahl's work makes clear, it would be foolish to assume that philosophy can go on unaffected by empirical research. But that does not mean that philosophy is one with the sciences. Philosophy rather seems to begin when we are no longer certain about what to make of our world, or of our empirical data, or when our ordinary language comes into some kind of conflict with the languages of the sciences, or when we feel that the words we used to be able to lead now seem to resist our guidance, or when some of our fundamental presuppositions of the order of things are challenged.

One might say that the essays in this book, in different ways and for varying reasons, echo Wittgenstein's effort to take philosophical language back to the everyday.[4] What that kind of philosophical movement attempts to accomplish is not the development of an alternative theory – with our ordinary language use employed as a standard of correctness – with which to counter other theories. If there is a quarrel with alternative 'theories' here, it is not about whether or not it is philosophically 'kosher' to depart from the ordinary, nor about presenting alternative solutions to the theories' problems, but about what our ordinary world is like. For example, Segerdahl's principal objection to particular forms of empirical research is not that the linguistic skills of apes ought not to be tested; rather, it is that such research typically operates with a dubious conception of what linguistic skills are. They are testing a picture of language, a theoretical model, and not our actual language – which is far more complex and multifarious than the picture suggests. Olli Lagerspetz's essay 'Rape among Scorpionflies, Spouse Abuse among the *Mantis Religiosa*, and Other "Reproductive Strategies" in the Animal and Human World' discusses similar problems in relation to evolutionary psychology. With Peter Winch as conversation partner, Lagerspetz presents a subtle form of Wittgenstein-inspired criticism of scientism – emphasizing the weight and importance of our ordinary language and the difficulties involved in simply disconnecting an everyday psychological (and evaluative) notion from its 'ordinary' home – without simply 'stating' that we go wrong if we don't talk the way we ordinarily do.

If we return to my first example with these remarks in mind, we may now see that what initially looked like something that naturally lends itself to philosophical

---

[4] Wittgenstein, *Philosophical Investigations*, §116.

reasoning, no longer easily does so. Above, I said that the first example *perhaps* is the easiest for contemporary academic philosophy to adopt since it may seem as if there is one single action portrayed there for us to pass judgement upon. I also said that the fictional example, rather than the anecdotal, invites a kind of generalization that more personal stories *appear* not to. My emphasis on 'perhaps' and 'appears' is meant to underline that such renditions of the example would be rushed. It is true that we get a vivid and colourful illustration of one action in the film sequence. We do see a man 'dropping' his knife in such a way that it penetrates the leg of a man who has no feeling in his legs. But is that all we see? And why do we see *it*? And what does 'seeing it' mean in this context? And is the nature of the action something we can understand without also considering the fact that it occurs as a scene in a movie?

If we ask what work this scene does in the film – which is not necessarily the same question as asking what work its author, the director, intended it to do – then we must also focus on the fact that it serves the function of introducing characters, Johner (the 'brute') in particular. This is where we learn what kind of a man he is: hard, strong and insensitive; a force to be reckoned with. We learn this by observing what he does and how he responds to others. And one might also say that when he, as a response to Call destroying his knife, tells us that 'I am not a man with whom to fuck', he is saying something we knew already, adding merely that we should also understand him to be, as it were, a man of principles. Thus, it would be a simplification to say that there is a single action there for us to pass judgement on. If 'dropping' a knife in the leg of a disabled man was a 'neutral fact' for us to pass judgement upon, the scene wouldn't work.

One might put it this way: the scene 'works' – that is, it manages to introduce this man in a convincing way as *this* kind of man – *because* we do not restrict the realm of the moral to questions of pain and suffering alone; *because* we do not first see one action and then go on to query whether it was morally blameworthy or praiseworthy; *because* we do not see one action in isolation. Indeed, one might say that even this example should lead us to the thought that philosophical understanding of morality requires a widened sense of the moral, that sensitivity towards context is indispensable and that 'mere calculation' is not the procedure we normally employ when we assess one another, that moral understanding is not to be reached if we focus solely on actions and choices. In fact, one way in which we learn that Johner is a brute comes from the fact that he does, as it were, calculate. Johner gives us 'rational' reasons for his actions that are supposed to trump the others' objections to his behaviour: Vriees has no feeling in his legs *therefore* it is not morally blameworthy to let a knife penetrate his leg. Perhaps one might even say that Johner's being presented as calculative is internal to the presentation of him as brutish. Call, Vriess and, I suspect, most of the spectators of the movie respond by saying 'no!' on no other basis than that this way of relating to another person is disrespectful, undignified.

We should also note that the way Johner is introduced *as* a brute is by linking him to animals. Johner's phony monkey laughter can be seen as a way of presenting him as 'not fully human'. Again, it is fairly unimportant whether or not the director had that intention. What matters is that it comes natural to us to think that a brutish human being is more 'animal' than the rest of us. In that sense, the 'film thinks', as Stanley Cavell has put it, by means of holding up a picture in front of us for us to relate to, regardless of what its author intended.[5] The animality of the scene was there all along, making it tick, even though we scarcely noticed it. And the mere fact that the scene works makes it clear to us (if we reflect upon it), that it comes natural to us to say that *one* contrast to the human is 'the animal'. If we, without hesitation, learn that *this* man is brutish partly on the basis that he seems to be more of an animal than the rest of us, then this picture of the animal is something that we all, to some extent, share. It is deeply rooted in our culture, our form of life.

Similarly, it is no wonder that we become baffled when an ape is rebuking a human being. The talking ape disrupts the order of things, makes some of our most fundamental beliefs visible to us and forces us to look at them anew. Segerdahl's disruptive experience of being reprimanded by a bonobo, is thus not merely a description of how faulty our pictures of (some) animals are, but also a testimony to the ways in which they are different from us. The fact that a human is stunned by a talking ape means both that apes are more similar to us than we might have thought, *and* it is an acknowledgement of our separateness. This also shows that when Segerdahl talks about 'encultured apes' or about an 'ape/human culture', he is not denying that apes *are* different from us. The point is rather that preconceived views can be disturbed when made visible and that it often is hard, very hard, to elucidate and understand one's most fundamental beliefs and attitudes – those that, in Wittgenstein's sense, are not 'opinions'.[6]

Mikel Burley's 'Wittgenstein, Wonder and Attention to Animals' also bears witness to how our encounters with animal life can bring facts about our own lives, our understanding and our morality, clearly into view in a way that is difficult, or even impossible, to attain without such experiences. The recognition of the interplay between estrangement and familiarity that Burley describes functions to make visible facts about ourselves that we might have taken for granted or thought of as non-existing, or repressed from philosophical discourse on the assumption that they are irrelevant. Attention to animal forms of life functions as a contrast that brings about a clearer conception of our own humanity.

Furthermore, the attention to animals here paid, shows that philosophical clarity about these matters (about the human–animal relation, about the human's engagement with its others, about the nature of moral discourse and evaluation) must take the complexity of contextuality (linguistic *and* cultural) into account, and that we do not understand them if we focus solely on singular actions or judgements and statements and scrutinize them on their own. My three examples are partly meant to bring out the misguidedness of the attempt to single out individual actions, judgements or statements. I have tried to show that even that which might look like 'one action' with a clear sense – say, Johner's little knife-party – will be misrepresented and misunderstood if thought of only in terms of already existing philosophical theories. The death of an

---

5   See 'Stanley Cavell in Conversation with Andrew Klevan,' in *Film as Philosophy: Essays on Cinema after Wittgenstein after Cavell*, ed. Rupert Read and Jerry Goodenough (Basingstoke: Palgrave Macmillan, 2005), 186.
6   Wittgenstein, *Philosophical Investigations*, §242.

animal can show us that the tendency to intellectualize and to remove all emotional elements of moral understanding may give us something that is more easily dealt with by means of theoretical models, but it might be at the cost of misrepresentation. And that a human's moral language can alter to such an extent that a joyful dinner is turned into something utterly remote and incomprehensible shows that the forms of life we lead are not static but in motion.

Together we can see these points converging in a call for a more sensitive form of philosophical method. It is a call for a reversal of philosophical strategies. The world should not be recast so as to fit our theoretical ideals; our philosophizing must be recast so as to be in touch with our world. This is one reason why I wanted to start with examples, with stories about our world, rather than theoretical accounts.

3. Different forms of forms of life

The essays collected in this volume all draw on Wittgenstein's philosophy in various ways, some more indirectly than others. Now, even though Wittgenstein's writings include a fairly large number of references to animals and animal life – the most famous being the lion we could not understand even if it spoke to us[7] – it would be a mistake to say that Wittgenstein had his own philosophy of animal life or animal rights. This means that there is no unified conception examined here; and that is another reason why I wanted to introduce the themes under discussion here by means of descriptions of human–animal relations, fictional and real, rather than starting off from an elaboration of the recurring notions, phrases, references, themes.

But even though there is no 'philosophy of animal life' to be extrapolated from his works, animals and discussions of animals do play an important, and perhaps surprisingly large, role in Wittgenstein's thought. Nancy E. Baker's 'The Difficulty of Language: Wittgenstein on Animals and Humans' is, at one level, a thorough exposition of the large number of 'animal references' there are in his work. At a deeper level, it shows us why and how animals actually matter to Wittgenstein's thinking. In particular, Baker can help us see how Wittgenstein's focus on language and the ability to use language does not make his philosophy anthropocentric but offers us a fresh perspective on the human–animal relation. In particular, Baker shows how reflections on animal life includes a widened sense of what it is to have a language, and of how one's (animal) surrounding cannot be circumvented.

Wittgenstein's talk about different 'forms of life' – clearly one of the most frequently employed phrases of Wittgenstein's in this volume – should not be seen as a definitive and well-established theoretical notion that can easily be appropriated. 'Form of life' (*Lebensform*) is not a technical concept. It is tempting to summarize different occurrences of that phrase into a constructive or positive account, thereby downplaying

---

[7]   Ibid., p. 190. (Also discussed in Baker's 'The Difficulty of Language: Wittgenstein on Animals and Humans' and Burley's 'Wittgenstein, Wonder and Attention to Animals' in this volume.)

Wittgenstein's own methodological goal not to present theories or to do hypothetical philosophy.[8]

If there is a standard account of Wittgenstein's idea of forms of life, it is that the meanings of our words are communal and that sharing a language means sharing a worldview. This is not exactly false, but it easily leads us into questions concerning linguistic and cultural relativism on the one hand and to a specific charge of conservatism (that is often directed against so-called ordinary language philosophies) on the other. That is, if one assumes that we hold the views we hold simply because we share a language, then it may appear as if it is impossible for us to, as it were, think outside the box and that we cannot really change the way we perceive things. As a positive account of how the world is, a metaphysical thesis of sorts, the idea of 'forms of life' seems to be at once static (since we cannot choose which language to use) *and* ad hoc (since it may seem as if the way we perceive our world and our relations to each other is 'merely conventional').

This rather common understanding of 'forms of life' may lead us to think that the form of life we happen to be in sets the *standard* for what is right and what is wrong. *That* thought may in turn lead us in two opposing directions. Either we think that we are, as language users, subordinated to the law of language. Or we may come to think that we can more or less choose our own standard of correctness since language, and so our form of life, is conventional. Neither of these options seems right to me.

In relation to this, I think that Cavell put his finger on something immensely important when he noticed that Wittgenstein's famous remark about the difference between agreement 'in opinions' and in a 'form of life'[9] is easy to misunderstand if one does not also take Wittgenstein's equally recurrent talk about the natural into account.[10] Cavell's point is that we should not understand Wittgenstein's talk about forms of life in a conventional (contractual) sense. We agree and disagree; but these agreements and disagreements seem to have a common ground, or backdrop, against which they make sense. This 'backdrop' – and I think we need to take all metaphors attempting to capture this with a pinch of salt – does not consist of 'opinions' that 'we' have (somehow and indeed very mysteriously) agreed upon. That's exactly the picture that Wittgenstein is contesting. Rather, the contrast Wittgenstein draws between agreement of opinions and in a 'form of life' suggests that we agree *in* language (not *about* language). And we do not *choose* our language (as we may choose to employ particular words on particular occasions and rebuke or praise our fellow beings for employing others). It is also clear that we cannot choose what language to speak – 'clear' in the sense that it is unclear what it would mean to do so (if we leave out the obvious point that it might be possible, in some contexts, to choose between different natural languages like English, Swedish, French). We cannot choose just *any* words of our language to employ either. That is not because it would go against what we have

---

[8] Ibid., §109.

[9] Ibid., §241.

[10] Stanley Cavell, *This New yet Unapproachable America: Lectures after Emerson after Wittgenstein* (Alburquerque, NM: Living Bath Press, 1989), 41f.

agreed upon, a breach of contract as it were, but because we would make ourselves incomprehensible. Words belong in contexts, depend on them, but they do not belong in just *any* context. When I or you employ a word in a context – I or you employ it in a place, at a location in our world, in nature. These are not 'conventions'. Our form of life is social, and so is language. But our form of life is also nature, and so is language. 'Commanding, questioning, recounting, chatting, are as much a part of our natural history as walking, eating, drinking, playing.'[11] Thus, focusing merely on the social character of language, easily leads one to think of forms of life in a 'contractual sense', which downplays the dependence we have on nature.[12] But if we call this 'dependence', we must also add that it is an open-ended, renegotiable, dependence.

This means that it is only in a very broad sense that the notion of form of life captures something linguistic, even though Wittgenstein says that 'to imagine a language is to imagine a form of life'.[13] Wittgenstein here uses the notion of form of life negatively in order to debunk pictures of language according to which it is merely a system of signs. He is not using 'language' to describe the sense of form of life. Thus, the notion here comes in, not as a positive account, but as the perfectly ordinary phrase it is, and he reaches for it when trying to show that language is neither something cut off from the world nor something disconnected either from our relations to one another or from nature. Language is, in that sense, not something we employ. It is our habitat.

References to backgrounds or frames or backdrops for our judgements and choices may easily lead one into positing dubious metaphysical claims. But that need not be so. A background can also be hard to see, and thus easy to neglect when philosophizing, precisely because it is always *there*.[14] That which we take for granted, we take for granted. It is not doubted, not proved, not seen, not hidden. We are, as it were, comfortable in it; it is certain, not a piece of knowledge. These are the themes explored in Julia Hermann's essay 'Man as a Moral Animal'. Hermann takes this sentence from *On Certainty* as her starting point: 'I want to conceive it [this kind of certainty] as something that lies beyond being justified or unjustified; as it were, as something animal.'[15]

This paragraph of Wittgenstein's is preceded by two very interesting remarks:

> 357. One might say: 'I know' expresses *comfortable* certainty, not the certainty that is struggling.

> 358. Now I would like to regard this certainty, not as something akin to hastiness or superficiality, but as a form of life. (That is very badly expressed and probably badly thought as well.)

As Hermann makes clear, Wittgenstein perceives the platform of our knowing as something that includes many sentences we might want to call empirical. This means that even the empirical world – biological, physical, physiological, etc. – is part of the

[11] Wittgenstein, *Philosophical Investigations*, §25.
[12] Cf. Cavell, *This New yet Unapproachable America*, 41f.
[13] Wittgenstein, *Philosophical Investigations*, §19.
[14] Ibid., §§129, 415.
[15] Ludwig Wittgenstein, *On Certainty* (Oxford: Blackwell, 1979), §359.

ground on which we stand when we pass judgements or condemn or praise actions. Thus, Hermann enables us to see the connection between Wittgenstein's vision of language and a natural (as opposed to a merely conventional or contractual) sense of forms of life.

Interestingly, as Wittgenstein returns to the phrase 'form of life', he adds the parenthetical remark 'That is very badly expressed and probably badly thought as well.' Now, what is badly thought and expressed here?

When we say 'I know', we are comfortable in our certainty. But is there a similar question of comfortableness when it comes to one's form of life? It is probably true that one need not be comfortable in one's form of life. The feeling of not being at home in this world is not as unusual as one might want to think. But in such a case, I also think that the sense of discomfort is not simply the opposite of comfort. It is likely that the sense of estrangement will be noticed and experienced, whereas the comfortable sense of being at home in one's language will not. The meaning of comfort (and discomfort) seems to differ in relation to 'knowledge' and 'forms of life'. The certainty that Wittgenstein aims at elucidating here is the certainty that is 'beyond being justified or unjustified'. It is something that we *do not* think about (which is not to say that it is something that we *cannot* think about, re-evaluate or reconsider). We might say that what is badly expressed here is the implied suggestion that one can feel comfortable or uncomfortable in a form of life. It is something that surrounds us and guides us without reflection, tacitly, naturally. It is animal.

Hermann shows that as we form our views about the world, within a language, there are always elements, logical and empirical, that we do not consider. It is not that we cannot reconsider them after they have surfaced in our minds as something that *could* be up for grabs. Rather, we do not know *now* what it would mean to do so. One might say that the senses of our words and the role they play in our lives would have to change for such a question to arise. Such a question *can* arise, it's just that it doesn't. And there is seldom any sense in saying that it *should*. Hermann's main point is that this extends to questions of ethics. There are a great number of things that need to change if we are to start thinking of the sentence 'Killing is wrong' as *not* right.

Thus, we may talk about 'form of life' as a phrase employed to capture the fact that linguistic sense, and so also moral sense, is intertwined with the way we lead our lives culturally and socially *and* with and in relation to empirical 'facts'. This brings us back to the fate of Snow White and Sooty.

That example was at first brought into play in order to show that context matters. But I also said that the reference to the change in my friend's form of life could not be employed to *explain* what is right and what is wrong in relation to this example. Hermann can now help us see why. Explanation requires justification, and the way we lead our lives is not based on sentences whose sense we first grasp and then go on to ask whether we are justified to believe them or not.

My first comments on Sooty and Snow White included the claim that it matters that my friend and his sister 'were kids, that they were used to seeing animals being slaughtered and prepared for meals, that this particular meal was cooked by their father (a trustworthy and upright man) and so on'. We may see this as an expression of the form of life they *had*. I also said that it is important to note that my friend now finds this episode of his life near to incomprehensible. And this sense of astonishment, of wonder, may make us search for justifications for our outlooks on life (present and previous). So we may start looking for arguments that enable us to disentangle our confusion. We feel we need to establish the reasons behind the change. (This is right because . . . *That* was wrong because . . .) Now, the problem is that the arguments that are supposed to justify seem to come in at a level when the questions are already settled. They prove what is already established.[16]

One obvious difficulty here is that we may employ the same words in two different forms of life. This easily gives us the illusion that we are talking about the same thing. But as life shifts, our concepts change – they are 'turned' differently (not exchanged). We may say, with Cora Diamond, that '[a] pet is not something to eat, it is given a name, is let into our houses and may be spoken to in ways in which we do not normally speak to cows or squirrels'.[17] This is the way the concept of a pet is 'turned' – now. But this does not mean that life, and so our concepts, may not be turned differently. This is the sense of Diamond's stark formulation 'it is not "morally wrong" to eat our pets; people who ate their pets would not have pets in the same sense of the term'.[18] As Rami Gudovitch shows in 'What's Wrong with a Bite of Dog?', the way our concepts are turned is not dependent upon a general moral principle that can be scrutinized in an impartial manner, but on the senses of our expressions. And these, in turn, cannot be disconnected from the way they function in our lives.

It is important to note here that our lives can and do change. And so does our language. And if we now retain the 'biological' sense of our forms of life, we can see *how* it matters that a human's relation to animals is interconnected with nature, culture, environment. It is thus a mistake to think that we simply have two forms of life, two 'entities' as it were – human and animal – that are possible to describe in a neutral manner. The sense of estrangement from his former 'I' that my friend experiences is, as I see it, a result of the hardness involved in acknowledging conceptual change, in seeing how slowly and gradually our concepts change shape, and so how they (our concepts) change us. It is the idea that the words 'pet' or 'animal' or 'human' or 'thing' denote particular entities that simply *have* this or that property, and require a specific treatment in virtue of their having it, that is leading us astray here.

Anne Le Goff's 'Living with Animals, Living as an Animal' explores the thought that having a language is not a 'property' that simply determines the essence of a being, and that this supposed essence in turn grants 'objective' rights for a being to be or not to be treated in this or that way. Le Goff approaches this theme by means of a discussion of the similarities and differences between two perspectives influenced by Wittgenstein's thought: Cora Diamond's and John McDowell's. Starting off from the recognition, shared by Diamond and McDowell, that the human is also an animal, Le Goff goes on

---

16   Cf. Diamond, 'Anything but Argument?'
17   Diamond, 'Eating Meat and Eating People,' in *The Realistic Spirit*, 324. Also discussed in Rami Gudovitch's 'What's Wrong with a Bite of Dog?'
18   Diamond, 'Eating Meat and Eating People,' 323.

to compare what significance the recognition of this fact has. McDowell's point of view is that language and reason are uniquely human, which suggests that there is a radical divide between these two forms of animal existences. And since reason and language exist only on one side of this divide, McDowell's view entails that language may not be 'apt' to capture the form of animal life that is not human. Thus, philosophizing about human animals has little to gain from reflections about (mere) animals. Diamond, on the other hand, stresses that we cannot understand the human if we do not also understand our relation to animals. Reflection on animal life within philosophical thinking is thus, on this view, not only possible but necessary. But it may require that we are willing to call our own conceptions of philosophy and of language into question.

Of course, no one denies that there are differences between animals and human beings. The question is what we are to make of these differences. As Stefano Di Brisco argues, McDowell's emphasis on human uniqueness has been developed with a specific type of question in mind, and it has rendered questions about what we *are to make* of the differences obscure. This is one way in which we humans can learn more about ourselves, but it may require that we adopt the 'look and see' strategy that Wittgenstein advocated. That is, our relations to animal life can only be adequately approached from within the stream of life.

In 'Ethics and Language: What We Owe to Speakers', David Cockburn explores the philosophical search for grounds for features of our lives. In particular, he shows that the question of what it means for a creature to have a language is not adequately addressed if one thinks of it as something that should ground the fact that a creature has something *else* (consciousness, conscience, reason, a soul). Taking the other to be speaking (or not speaking) already involves an ethical 'stance'. The other demands something from you. The idea that there is, say, a specific number of words that the other needs to master in order to qualify as this or that, distorts the very features one wishes to investigate.

There is a sense in which it is an illusion to say that we control ethics, that we can reach full perspicuity of it and that we may choose our moral outlook on the basis of reason. The opposite seems nearer the truth: ethics controls us. This, in turn, moves the concept of ethics closer to that of logic. It is that which we reason through, not about. And as we investigate it, it is both the object of study and the subject studying. Just as we must use logic to elucidate the sense of logic, so we reason 'ethically' when we try to elucidate ethics. This is one of the reasons, I think, why Wittgenstein persistently stressed that *all* we can do in philosophy is to describe, and that the production of 'theories' will distort our perception. This makes ethics more difficult. Reason is under reason's investigation. It is philosophy.[19]

[19] My thanks to Lars Hertzberg, Mikel Burley, Nora Hämäläinen and the participants at the Research Seminar in Philosophy at Åbo Akademi University, for valuable comments to an earlier version of this text.

1

Humanizing Nonhumans: Ape Language Research as Critique of Metaphysics

Pär Segerdahl

1. Introduction

Suppose that a philosopher wrote copiously about 'the human': about human language, about human practices, about human culture. It seems natural to assume that this philosopher would presuppose a contrast between human and animal, and that he would be writing about what he considers uniquely human. When philosophers contrast human and nonhuman, the part of the nonhuman tends to be played by the animal.

I know of at least one philosopher, though, for whom the philosophical opposite of the human was not the animal. I am thinking of Friedrich Nietzsche, who wrote: 'Where *you* see ideal things, *I* see what is human.'[1]

Nietzsche did write extensively about the human (and the all-too-human), but it seems to me that the contrast he is making is rarely with the animal. With a few exceptions, he typically emphasizes the human in opposition to what, in a certain sense, is philosophical.[2] His critical efforts are directed against tendencies to idealize human reality as if it manifested abstract concepts. The opposite of the human, for Nietzsche, was the over-intellectualized version of reality that philosophers construct in what for him was their *failure* to deal with human realities. 'Ideal things' were the opposite of what is human; *they* were the nonhuman for Nietzsche, not the animals.

Ludwig Wittgenstein, I believe, gradually became another philosopher who saw human and nonhuman things in this Nietzschean manner. If *Philosophical Investigations* assumes an opposite of what is human, this nonhuman typically exists within us, in the form of intellectual tendencies that traditionally came to expression in

---

[1] Friedrich Nietzsche, *On the Genealogy of Morals and Ecce Homo*, trans. R. J. Hollingdale and Walter Kaufmann (New York: Vintage, 1969), 283.

[2] An exception occurs in the opening of Friedrich Nietzsche, *On the Advantage and Disadvantage of History for Life*, trans. Peter Preuss (Indianapolis: Hackett, 1980). A herd of grazing animals is distinguished from humans by their having no memory. They live without being burdened by history. The image of the herd is used to identify problems in contemporary culture, and I want to say that in Nietzsche's treatment of these problems, animals have vitalizing functions.

lofty idealizations of what it means to be human. Where Nietzsche complained about philosophers' tendency to place their highest concepts – 'the most general, the emptiest concepts, the last smoke of evaporating reality – in the beginning as the beginning'[3] – Wittgenstein suggested ways of bringing words 'back from their metaphysical to their everyday use'.[4] This everyday use Wittgenstein portrayed as taking place within human forms of life.

I want to inquire into this new philosophical significance of the human, where our nonhuman opposite no longer is identified with the animal, but with what Nietzsche called 'ideal things'. I want to ask what it means to compare human and animal *after* Nietzsche and Wittgenstein, when the nonhuman turns out to exist within us, as an idealizing tendency.

2. Comparative psychology as metaphysical endeavour

My discussion will be concretized by some unexpected findings in the experimental comparison of humans and animals. The findings I have in mind have not yet been understood fully, I believe, because they require that rare sensitivity to human and nonhuman things that I just mentioned.

The traditional philosophical obsession with defining what distinguishes humans from animals has today been taken over by a field with scientific ambitions called comparative psychology. Comparative psychologists differ from traditional philosophers in that they use experimental approaches and a perspective from evolutionary biology. But on one significant point they agree with the tradition. They take for granted that the difference between humans and other animals – the *first* difference with which all other differences are supposed to begin – is a grand hidden one: a defining trait having to do with the human intellect, or with 'the architecture of the human mind'.

Comparative psychologists motivate their efforts in the laboratory by first highlighting differences between humans and other animals that are open to view and do not require laboratory work. A recent contribution to the field is introduced with the following observation:

> Human animals – and no other – build fires and wheels, diagnose each other's illnesses, communicate using symbols, navigate with maps, risk their lives for ideals, collaborate with each other, explain the world in terms of hypothetical causes, punish strangers for breaking rules, imagine impossible scenarios, and teach each other how to do all of the above.[5]

---

[3] Friedrich Nietzsche, *Twilight of the Idols/The Anti-Christ*, trans. R. J. Hollingdale (London: Penguin, 1968), 37.

[4] Ludwig Wittgenstein, *Philosophical Investigations*, trans. G. E. M. Anscombe (Oxford: Blackwell, 1953), §116.

[5] Derek C. Penn et al., 'Darwin's Mistake: Explaining the Discontinuity between Human and Nonhuman Minds,' *Behavioral and Brain Sciences*, 31 (2008), 109.

The authors go on to suggest that these easily noted differences have a mental cause more difficult to discern; one not yet identified by science. All we can state with certainty at this early stage is that, 'human minds are qualitatively different from those of every other animal on the planet.'[6] The aim of performing laboratory tests with chimpanzees and human children is to tease out the invisible qualitative mental difference that explains the noticed differences: the difference that is placed at the beginning *as* the beginning. In the article just quoted, this primary difference is suggested to consist in 'our species' unique ability to approximate the higher-order relational capacities of a physical symbol system.'[7] As the wording suggests, the spirit has become more technical than it is in most philosophy. Philosophers traditionally talk about grander things, such as 'reason'. Moreover, a demand has been added: the discussion must proceed on the basis of empirical evidence from the laboratory. Still, the overall intellectual goal is the same as in the philosophical tradition: to find the underlying mental difference between humans and the other animals that explains the noticeable behavioural differences.

The comparative-psychology approach to the comparison of humans and animals has two features that I want to emphasize before moving on. The first is that despite the attempted evolutionary perspective, the opposite of the human, the nonhuman, is the animal (or 'every other animal on the planet'). The second feature is that the primary difference, the one that is supposed to ultimately distinguish the human (the one with which all other differences begin), is hidden, is abstract – precisely what for Nietzsche was the nonhuman.

The forms of life which Wittgenstein invoked to humanize metaphysical words are for comparative psychologists merely a preliminary starting point: indirect evidence pointing towards underlying psychological realities that can be uncovered only by bringing children and chimpanzees into the laboratory and comparing their performance on cleverly designed psychological test tasks. Comparative psychology makes a virtue of neglecting the following kind of observation:

> The aspects of things that are most important for us are hidden because of their simplicity and familiarity. (One is unable to notice something – because it is always before one's eyes.)[8]

The question is: how can we compare humans and animals without overlooking such an observation? How can we compare humans and animals while identifying our *common* opposite as those intellectual tendencies that Nietzsche and Wittgenstein saw as leading us to nonhuman chimeras?

3. Human-enculturated apes

Comparative psychology does not have to be practiced in the metaphysical spirit that I just sketched. Although they are a minority of a minority, there are comparative

---

6   Ibid.
7   Ibid., 111.
8   Wittgenstein, *Philosophical Investigations*, §129.

psychologists whose experimental work does not have the ultimate aim of uncovering a primary mental difference with which all other differences between humans and animals allegedly begin.

Suppose that what Nietzsche and Wittgenstein saw as human things could make a difference; that they could change living beings. What happens if apes are initiated into human forms of life, being raised by human parents? Could such rearing be the beginning of hitherto unseen relationships between humans and apes? Can rearing be the beginning of new similarities? These were the questions the psychologists I have in mind asked. They too took apes and children into the lab and compared their performance on psychological test tasks. But *before* they brought the apes into the lab, they devoted many years of their own and the apes' lives to sharing daily existence down to the most trivial detail, taking the same responsibility for the apes as parents do for their children.[9] What happens when these humanly reared apes are brought into the lab and compared with children with analogous background, or with apes reared in other ways? More importantly, what happens long before the apes are brought into the lab, during the many years of daily coexistence with humans? Into what kind of 'animal' does this initiation into the human turn these apes?

The apes with whom I am concerned are described in the literature as human-enculturated apes, or simply as enculturated apes.[10] In the most interesting cases, they learn things that animals are supposed not to be able to do, such as communicating in language at the level of a 2½-year-old human child, pointing declaratively, manufacturing and using their own stone tools, understanding what another believes is the case and many other human practices and ways of life.[11]

I want to draw attention to two philosophically significant features of the human-enculturated apes. The first is that these apes alter the landscape of noticeable differences that comparative psychology typically treats as a given fact and as its motivation for doing more revelatory work in the lab. After millennia of philosophical discussion, the basic facts that we thought we could take for granted about what animals do and do not do have changed, because of the recent appearance of enculturated apes.

---

[9] See, for example, William M. Fields et al., 'The Material Practices of Ape Language Research,' in *The Cambridge Handbook of Sociocultural Psychology*, ed. Jaan Valsiner and Alberto Rosa (Cambridge: Cambridge University Press, 2007).

[10] See, for example, Michael Tomasello, *The Cultural Origins of Human Cognition* (Cambridge, MA: Harvard University Press, 1999), 34–6.

[11] For data on language comprehension, see E. Sue Savage-Rumbaugh et al., *Language Comprehension in Ape and Child*, *Monographs of the Society for Research in Child Development*, 58(3–4) (1993). For data on apes' ability to participate in conversational exchanges, see Janni Pedersen and William M. Fields, 'Aspects of Repetition in Bonobo–Human Conversation: Creating Cohesion in a Conversation Between Species,' *Integrative Psychological and Behavioral Science*, 43 (2009), 22–41. For data on pointing, see Janni Pedersen et al., 'Why Apes Point: Indexical Pointing in Spontaneous Conversation of Language-Competent Pan/Homo Bonobos,' in *Primatology: Theories, Methods and Research*, ed. Emil Potocki and Juliusz Krasiński (Hauppauge, NY: Nova Science Publishers, 2011). For data on stone tool manufacture, see Nicholas Toth et al., 'Pan the Tool-Maker: Investigations into Stone Tool-Making and Tool-Using Capabilities of a Bonobo (*Pan paniscus*),' *Journal of Archeological Science*, 20 (1993), 81–91, and Nicholas Toth et al., 'A Comparative Study of the Stone Tool-Making Skills of *Pan, Australopithecus*, and *Homo sapiens*,' in *The Oldowan: Case Studies into the Earliest Stone Age*, ed. Nicholas Toth and Kathy Schick (Bloomington, IN: CRAFT Press, 2003). Apes' understanding of others' mental states is featured in the documentary, *Kanzi II.*

The second feature is that this transformation of the apes is brought about on the level of day-to-day living, far removed from all purported 'primary' mental causes.

The transformation that enculturated apes undergo occurs when they are embedded in the kind of circumstances to which Wittgenstein brought metaphysical words. It is remarkable that these forms of life, always before our eyes and therefore easily unnoticed, have humanizing effects in both cases. Words regain their familiar (but easily neglected) human forms and apes develop forms they did not have before.

4. The 'Nim prejudice' about enculturation

I am searching for a new outlook on the human–animal relation. If philosophy changes direction when we, like Nietzsche and Wittgenstein, identify the nonhuman with those 'ideal things' that philosophers previously saw as the essence of human reality (and contrasted with animal existence), then it is high time that we learned to compare humans and animals in correspondingly new ways. I believe that such new comparisons emerge from the work with enculturated apes.

In order to clarify these new comparisons, I need first to address the most widespread prejudice about the human-enculturated apes. Enculturated apes typically belong to ape language projects that investigate whether apes can be taught parts of human language.[12] The prejudice about these apes is the following: apes in language research must be specially trained to use linguistic symbols. They must be compelled to learn from humans through clever animal training techniques. Expressed another way, the prejudice is that young apes cannot respond spontaneously to long-term human contact in ways that are comparable to how human children respond to their long-term contact with human caregivers (parents).

This prejudice is so tenacious that most ape language researchers actually did train the apes they worked with. We are dealing with a prejudice so powerful that it governs not only public perceptions of ape language but also the research itself. That is what I meant by saying that the comparative psychologists whose work I want to use to concretize a philosophical turn are a minority of a minority. There are only a handful of ape language researchers in the world, and among these, only one research group consistently avoided training. And they were the ones with the unexpected findings that enable new ways of comparing humans and animals.

---

[12] Some landmarks in ape language research are W. N. Kellogg and L. A. Kellogg, *The Ape and the Child: A Comparative Study of the Environmental Influence upon Early Behavior* (New York and London: Hafner Publishing Co., 1933); Keith J. Hayes and Catherine Hayes, 'The Intellectual Development of a Home-Raised Chimpanzee,' *Proceedings of the American Philosophical Society*, 95 (1951), 105–9; R. Allen Gardner and Beatrice T. Gardner, 'Teaching Sign Language to a Chimpanzee,' *Science*, 165 (1969), 664–72; David Premack, 'Language in Chimpanzee?,' *Science*, 172 (1971), 808–22; Duane M. Rumbaugh, *Language Learning by a Chimpanzee: The LANA Project* (New York: Academic Press, 1977); Herbert S. Terrace, *Nim* (New York: Knopf, 1979); E. Sue Savage-Rumbaugh, *Ape Language* (New York: Columbia University Press, 1986); Savage-Rumbaugh et al., *Language Comprehension in Ape and Child*.

Let me illuminate the prejudice with an example: the experimental psychologist Herb Terrace's efforts to teach American Sign Language to a young chimpanzee named Nim Chimpsky.[13] Terrace's work built on the apparently obvious assumption that if you want to study whether it is possible to talk with an ape, you must first teach the ape a language in which you can attempt to communicate with it. For this primary educational purpose, Terrace designed a bare and small classroom where 60 teachers alternated trying to make Nim form signs with his hands. The purpose of the bare classroom was to avoid distracting Nim with things that young apes find more exciting than sitting still on the ground imitating how a teacher shapes her hands. Only when Nim's ability to form signs properly with his hands had developed could one see if he would use these signs creatively to communicate with humans. The project was considered by Terrace himself to have failed. When Nim's ability to form signs for various objects and actions developed, it seems he did not use them to communicate. He typically continued to do what he did in the classroom. When the human made certain signs, Nim made the same signs, mirroring the teacher's hand movements. That is not communication.

Observe that Project Nim applied the same metaphysical thinking against which Nietzsche and Wittgenstein were in rebellion. An idealization of human language as vocabulary and grammar was placed at the beginning of speech, as its foundation. The assumption was that in order to speak with an ape, you must first leave life behind as a distraction and move into the more abstract domain of first mental causes – a domain which, I repeat, was the nonhuman for Nietzsche and Wittgenstein. Needless to say, compelling apes to move into this nonhuman domain fails to humanize them. Terrace reported the failure in a famous article in *Science*.[14]

5. Young bonobo Kanzi and his human mother Sue

Significant variation is valuable in science and in philosophy, especially if the variation surprises you. A few years after Terrace published his negative results with Nim, another ape language researcher, Sue Savage-Rumbaugh, was training a wild-caught adult bonobo, Matata, to use so-called lexigrams: printed word symbols to which the ape points. The attempt to train Matata to use lexigrams was possibly even less successful than the work with Nim. However, while Matata, to no effect, was being taught lexigrams, another little creature was present. Matata's adopted son, Kanzi, was playing around them. Since he was considered too young to sit still on the ground and participate in a training programme, no one tried to teach Kanzi lexigrams, and sure enough, he constantly interfered with Matata's training and was charmingly annoying.

One day when Matata temporarily was taken away for breeding purposes and young Kanzi found himself alone with humans, to everyone's surprise he approached

---

[13] Terrace, *Nim*.
[14] H. S. Terrace et al., 'Can an Ape Create a Sentence?', *Science*, 206 (1979), 891–902.

the keyboard and, on his own initiative, began to point to various word symbols, like BANANA, JUICE, OUTDOORS and SWING. It was not evident what this spontaneous pointing to symbols meant. Was he talking? Was he asking for juice? Was he suggesting going outdoors? It was not entirely clear what was going on, but when Kanzi pointed CHASE and ran away with a tantalizing look on his face, it changed Sue Savage-Rumbaugh's relation to him and to what it meant to 'enculturate' an ape.[15]

Without having been specially trained, Kanzi seemed to have become someone who could face a person and say: *chase me.* That, at least, was how Sue responded to him, for she did chase him. By responding thus to Kanzi, and by doing so consistently, Sue Savage-Rumbaugh reversed her approach to ape language and to the comparison between humans and animals. She reversed her notion of what comes first and what comes last in matters of language and speech. She reversed her identification of what is distracting and what is essential in human language.

> Thus, I decided to abandon all instruction and focus my attention instead on what was *said to Kanzi* rather than on what we could teach him to say.[16]

Instead of locking Kanzi up in a classroom to avoid distractions, Sue connected to Kanzi and to his many ways of being playful, curious, afraid, sad or happy in the circumstances of a life that she began to share with him. Instead of forcing Kanzi into a meticulously planned training programme, Sue and Kanzi went out into the 55-acre forest surrounding the laboratory and exposed themselves to the unexpected events of a dense forest. Shelters were built where Kanzi and Sue, and a select group of (to Kanzi) familiar humans and apes (and dogs) accompanying them, could stop, eat and play. Each place was given an English name, such as 'Lookout Point', and a corresponding lexigram on the portable keyboard. Days and even nights were spent travelling in the forest, talking about where to go, what to eat, or, perhaps, the snakes or wild dogs that surprised them among the trees. When Kanzi used lexigrams, when he pointed FIRE, it was not a linguistic training practice initiated by Sue in order to teach Kanzi a language in which he might speak to her in the future. It was instant speech by Kanzi addressed to Sue; it was Kanzi suggesting making a fire outside one of the shelters; it was ongoing ape–human life, functioning as the beginning of changes up to then unseen in an ape.

Perhaps the most succinct description of the novelty of Sue Savage-Rumbaugh's approach to enculturation is the following: for Matata or Nim an excursion into a forest would have been a temporary relaxation from the enculturation process, as when a student is allowed to take a break from her linguistic studies. For Kanzi,

---

[15]   For more detailed accounts and discussions of this event in ape language research, see Savage-Rumbaugh et al., *Language Comprehension in Ape and Child*; Sue Savage-Rumbaugh and Roger Lewin, *Kanzi: The Ape at the Brink of the Human Mind* (New York: Wiley, 1994); Sue Savage-Rumbaugh et al., *Apes, Language and the Human Mind* (Oxford and New York: Oxford University Press, 1998); Pär Segerdahl et al., *Kanzi's Primal Language: The Cultural Initiation of Primates into Language* (Basingstoke: Palgrave Macmillan, 2005).

[16]   Savage-Rumbaugh et al., *Apes, Language and the Human Mind*, 26–7.

however, going to Lookout Point *was* how he learned to use the name of this place. In his case, it was impossible to draw a demarcation line between when he was being enculturated into language and when he was relaxing. His situation was analogous to that of a human child. We do not say of a two-year-old child, 'She has been practicing language for two hours now, she needs a break; let's go outdoors.' It's unclear what 'taking a break from language' could mean in the case of a child who is becoming a speaker for the first time. 'Taking a break' has clear meaning when older children study foreign languages. But then they already have become speakers in more boundless ways.

Metaphysical attitudes to what is human and nonhuman affect apes in psychological research. If you idealize vocabulary and grammar as the first mental cause of speech, as the primary mental difference that makes humans rather than animals into speakers, then if you want to see if an ape can learn to speak, you will begin by trying to place this assumed mental architecture in the ape's head as the beginning of its future speech. Such attempts have failed. If, on the other hand, you reverse your notion of what comes first and what comes last as Nietzsche and Wittgenstein did in philosophy, then you will from the outset share your existence with the ape, down to the most trivial detail, raising the ape as if you were its human parent. By thus in practice identifying 'the human' with the circumstances of daily life, rather than with some idealized abstract first cause, this primal exposure to human ways of life humanizes the ape. The ape becomes an enculturated ape, genuinely engaged with what hitherto was considered uniquely human.

6. Wittgenstein's children

In *Zettel*, Wittgenstein discusses several examples of children learning new words. He then asks himself, 'Am I doing child psychology?' He answers, no; he is rather making a connection between the concept of teaching and the concept of meaning.[17] If I understand him right, Wittgenstein considers children learning new words because such examples so clearly display the human circumstances of meaning: how the meaning of a word is not a 'first mental cause' of its use. The question, 'How can a child learn to use a word such as "think"?' emphasizes what has to be in place in the child's *life* in order for it to start using the new word. Wittgenstein's children help us see how seemingly abstract notions, like that of thinking, have their abode in human life; how they are integrated into the circumstances of human life ways. They are not pure ideas detached from these circumstances, causing their own use among them.

In *Zettel*, Wittgenstein made just this point:

114. One learns the word "think", i.e. its use, under certain circumstances, which, however, one does not learn to describe.

---

[17] Ludwig Wittgenstein, *Zettel*, trans. G. E. M. Anscombe, 2nd edn (Oxford: Blackwell, 1981), §412.

115. But I *can teach* a person the use of the word! For a description of those circumstances is not needed for that.

116. I just teach him the word *under particular circumstances*.[18]

These remarks characterize what it means to learn to use words *for the first time*. It has to occur 'under particular circumstances'. A nine-year-old child who *already* talks about thinking, however, can learn the word for 'think' in a foreign language without the teacher having to care one bit about the human circumstances that Wittgenstein remarks are so central when a child learns to use the word for the first time.

Here Nietzsche's and Wittgenstein's critique of metaphysics becomes relevant not only for child psychology but above all for ape language research. Child psychology is done with children who are enculturated with their human parents. Even the most metaphysical comparative psychologist works with children whose upbringing is independent of his notions of what comes first and what comes last. Whether the psychologist is sensitive to the critique of metaphysics or not will not make a difference to his research subjects. Ape language research is fundamentally different since the apes are held in captivity. They are in the hands of the same psychologists who test them in the laboratory, and who may neglect the importance of teaching words in the right circumstances; who may believe that they can, or even *must*, teach language to apes in bare classrooms where life is treated as a distraction.

Ape language research is perhaps the closest we get to a 'philosophical experiment'. At least, it is an experiment that poses philosophical challenges for psychology. If all ape language researchers failed, if no ape learned to talk, then traditional thinking could prevail and one could continue thinking that apes simply do not have the mental architecture required as the first cause of speech. But there is significant variation in ape language research. Researchers who train apes in accordance with traditional ideas about first mental causes fail. The only researchers who succeed are those who rear apes as their human parents; who talk with the apes day and night, regardless of whether the apes initially understand, thus gradually initiating them into the circumstances where new words can be spoken and learned. These researchers begin with what I believe we need to start acknowledging *as* the beginning.

7. Savage-Rumbaugh's apes

Kanzi is not the only ape who developed language in the circumstances of his life; he is just the first ape who learned to talk as Wittgenstein's children do. Later, Kanzi's half-sister, Panbanisha, her son, Nyota, and a common chimpanzee, Panzee, developed language the same way Kanzi did: by being reared by human parents who waited for circumstances where new words could be genuinely spoken and learned.

If you want to study the ambitious laboratory test and the data that support the claim that Kanzi understands spoken language at the level of a 2½-year-old child, the

[18]   Ibid., §§114–16.

book to read is *Language Comprehension in Ape and Child*.[19] In the test that is reported and discussed in this monograph, Kanzi is asked to do several hundred more or less novel things that he had never been asked to do before, like 'Can you hug the ball?' or 'Go to the colony room and get the phone' – and he responds appropriately to these and most other commands, just hearing them spoken.

It is easy to be impressed by Kanzi's performance in the test. But although the data concerning his responses to about 650 novel sentences reveal more than you would expect of his bi-species ways of living, they do not reveal how Kanzi originally *acquired* the language he demonstrates in the test. When a child demonstrates the same mastery of language, it is easy to assume that the child has the right mental architecture causing its behaviour, namely: an appropriate range of vocabulary and a capacity to understand recursive grammar. When Kanzi takes the test as successfully as a 2½-year-old child, it is tempting to think that training produced a similar mental architecture in Kanzi's head, causing *his* behaviour. That is also how the test is discussed critically: does it really prove grammatical competence in Kanzi?

The significant fact, however, is that Kanzi was not trained. Apes who are trained according to the notion of vocabulary and grammar as the first mental cause of speech fail the test. What is most important *precedes* the test and has to do with Kanzi's rearing; with his *Zettel*-like enculturation into language; with the *circumstances* under which he learned new forms of language. Kanzi's way of *acquiring* the language he demonstrates in the test ought to dissuade us from looking for grammar as the first mental cause of his test performance.

8. How an ape learns to use the word 'monster'

I first visited Kanzi and his fellow humans and bonobos (and dogs) in August 2001. On the first day of my visit, I was told by one of the researchers, William Fields, the human parent of Nyota, to just sit still on a chair and observe the apes. However, a previously employed caretaker came by to visit the apes and she wanted a keyboard to talk with Panbanisha. I forgot my promise to just sit and observe, got up from the chair and tried to explain in broken English where I had seen a keyboard. However, Panbanisha already had a keyboard inside the enclosure where she was, and she put her finger on one of the lexigrams and looked dissatisfied. I asked the caretaker what Panbanisha was saying. She studied the keyboard, blushed and told me that Panbanisha was saying QUIET on the keyboard. A little later I was playing with Panbanisha's two sons. We were playing peek-a-boo through a window. Beneath the window there was a hole in the wall, and one of the bonobos, Nathan, stretched his arm through the hole. I looked at his hand, and although I was not allowed to, I could not resist the temptation to touch it. Nathan immediately withdrew his hand

[19] The language comprehension test with Kanzi is recreated in the video documentary, *Bonobo People*.

and ran out to mother Panbanisha, who recently had told me to be quiet. After just a few seconds, she burst into the room, trailing the keyboard behind her. She ran up to the window behind which I sat and hit it with her fist. She then sat down in front of me and put her finger on one of the lexigrams. The human father of Nyota asked her if she wanted to communicate with Pär, to which she responded with the short high-pitched vocalization that she and Kanzi use to answer in the affirmative.[20] It took some time for me to see which lexigram she pointed to, but she kept her finger firmly on the lexigram. Not until I shouted to Nyota's human father, 'She is calling me a monster,' did she remove her finger from the MONSTER symbol. She had spoken, and I had understood, and I sat there ashamed and overwhelmed by being rebuked thus eloquently by an ape.

How astonishing it is that an ape can use the word 'monster'. How would you go about teaching an ape the word 'monster'? How on earth did Panbanisha learn to talk about monsters? Well, no one had planned it. Gradually, the bi-species ape/human culture became such that there was a place in it for the notion of a monster. The circumstances appeared where this new word could be genuinely spoken. Sue Savage-Rumbaugh wanted to scare the apes from climbing up onto a roof where there were high-voltage cables, so she had one of the caretakers dress up in a gorilla suit and run around on the roof. This terrifying figure was sometimes described as 'the gorilla' and sometimes as 'the monster', and got two lexigrams on the keyboard. But just like children, the apes were fascinated by the scary monster and going out searching for it became a game. Sometimes caretakers even make monster movies for the apes. In these videos one caretaker dresses up as the monster while the others act as themselves, discovering the monster and driving it away. The apes then request to see these thrillers by pointing to the lexigram, TV-TAPE. Panbanisha is especially keen to explore her own fear by watching these movies, while Kanzi does not like to be scared and consistently chooses rabbit movies instead, where one of the caretakers dresses up as a harmless figure named BUNNY, who always brings presents.

Sometimes, the apes themselves dress up as monsters. They put on a monster mask or pull a blanket over their head, and then chase the other apes. If bonobos never became afraid of gorilla-like characters, if they were not fascinated by their own fear and never explored it through creative play, then the circumstances would hardly occur in the bi-species culture where the apes could learn the word 'monster'.

Wittgenstein could have used this ape example in the remarks that later became *Zettel*, had it then existed. Conversely, Sue Savage-Rumbaugh could have said, 'I just teach him the word *under particular circumstances*', as a comment on her life and work with bonobos – Am *I* doing ape language research? I am making a connection between the concept of enculturation and the concept of language.[21]

---

[20] For data on the bonobos' use of their voices in communication with humans, see Jared P. Taglialatela et al., 'Vocal Production by a Language-Competent *Pan paniscus*', *International Journal of Primatology*, 24 (2003), 1–17.

[21] See Segerdahl et al., *Kanzi's Primal Language*, for more details about this connection.

9. Home taking precedence over lab

Ape language research is often accused of being a human-centred practice unfairly judging apes on *our* terms. Because it imposes what is considered uniquely human upon nonhumans, the primatologist Frans de Waal views the research as evolutionarily misguided:

> Personally, I must admit to mixed feelings about ape language research. On the one hand, I see it as a thoroughly anthropocentric enterprise. A communication system for which evolution has specifically hardwired us (and perhaps only us: our brains are three times larger than the average ape brain) is being imposed upon another creature to see how far it can go.[22]

This widespread notion of ape language research as an anthropocentric enterprise does not consider the *Zettel*-like intricacies of the work that I am trying to communicate here. De Waal's critique assumes that language is the first mental cause of speech; that language is a system of vocabulary and grammar encoded in human brains. What ape language researchers do, it is claimed, is attempt to humanize apes by squeezing this mental architecture into ape brains that were never hardwired for it. Therefore, a common critique is that even if these 'trained subjects' do appear to speak and respond appropriately to speech, it cannot be *real* language. They only behave *as if* they had language, since their behaviour cannot have the right primary cause: they are 'aping' language.[23]

The critique that ape language research is anthropocentric gives expression to the same metaphysical mode of thinking that successful ape language research avoids. What ape language researchers are accused of doing is squeezing into ape heads precisely what for Nietzsche and Wittgenstein was the nonhuman: the last smoke of evaporating reality, as if it were its fresh beginning! These alleged 'human' terms are as misguided in their application to humans as they are in their application to apes. The example of Kanzi helps us see that the terms are intellectual constructions rather than facts of human biology. Studying in concrete detail how the notions of speaking, of asking and of answering become applicable to Kanzi during his enculturation helps us humanize our understanding of the concept of language.[24]

Successful ape language researchers like Sue Savage-Rumbaugh *do* humanize apes to the point where the apes will call you a monster if you misbehave. But they do it by beginning at the beginning; by beginning in the right circumstances; by beginning with the right rearing. Thereby, they uncover a human that our metaphysical self-image so far has prevented us from seeing.

I understand that it could be objected that any attempt to humanize apes inevitably is anthropocentric, precisely as de Waal claims. In a sense it is, of course, but since this

[22] Frans de Waal and Frans Lanting, *Bonobo: The Forgotten Ape* (Berkeley, CA: University of California Press, 1997), 44.
[23] See, for example, Joel Wallman, *Aping Language* (Cambridge: Cambridge University Press, 1992).
[24] This is what we attempt to do in Segerdahl et al., *Kanzi's Primal Language*.

particular attempt reveals *unexpected* human dimensions of the terms that we use for what is human, it *challenges* what de Waal assumes is being 'imposed upon another creature'. What Sue did was *living with* Kanzi. If you want to see *living with* an animal as anthropocentric, then this form of anthropocentrism challenges the more prevalent forms of anthropocentrism; for instance, the one that de Waal takes for granted when he discusses ape language research. Moreover, as a result of sharing everyday life with apes, even the humans involved in the research changed.[25]

I want to suggest a more constructive response to the critique that ape language research is anthropocentric. I want to explain how I believe the research, when it successfully humanizes apes, works and how children and apes need to start being compared. The cultural psychologist, Jerome Bruner, once wrote that 'you could only study language acquisition at home, *in vivo*, not in the lab, *in vitro*'.[26] What Bruner meant, I take it, is that you can bring children into the lab; you can test them and study their developing language skills. But the lab is not, and cannot be, the place where children originally acquire the studied skills. Language acquisition must take place elsewhere, in the right circumstances. It must occur 'at home', where home is not only the child's private home, but also the home of our common language. One could call this home human culture, or the human forms of life, as these emerge in children's normal rearing in tandem with their developing language.

What Bruner said could be expressed thus: *psychological research with children inevitably exhibits home/lab duality*. You can use psychological experiments in the lab to measure children's psychological development, but the dynamics of this development does not occur in the lab, but at home, within the circumstances of human life. Without a home, no language worth measuring in the lab would develop.

Consider again Herb Terrace's chimpanzee, Nim Chimpsky: what did Nim's life as a test subject in psychology look like? Did he have a home where he could develop the language that later would be measured experimentally? Although he spent his spare time in a New York household, the place where Nim was supposed to acquire new linguistic signs was the tiny classroom, designed specifically to *avoid* what we call 'home' as if it were a distraction from language. In contrast to work with children, then, the research with Nim did not consistently utilize home/lab duality as a way of achieving relevant psychological results and valid comparisons between human and ape. Human test subjects display home/lab duality. The ape subject, Nim, was prevented from doing so. Terrace's way of comparing humans and animals seems unfairly asymmetrical.

Consider now Kanzi and Panbanisha. They were thoroughly tested over the years in controlled psychological experiments. But as I indicated with their acquisition of the word 'monster', how they originally developed language is distinct from how they were later tested in lab conditions. Despite the fact that Kanzi and Panbanisha are captive apes at a centre for psychological research, the people working with the apes created

---

[25] For a more thoroughgoing discussion of anthropocentrism and anthropomorphism in the work with enculturated apes, see Segerdahl et al., *Kanzi's Primal Language*, 109–17.

[26] Jerome Bruner, *Child's Talk: Learning to Use Language* (New York and London: Norton, 1983), 9.

home/lab duality for them and utilized this duality to achieve previously unimaginable comparisons between humans and apes.[27]

Here is an illustration of how this duality can reveal itself even within lab work. When Kanzi and Panbanisha enter the lab as test subjects the experimenter talks with them as familiarly as with a child in a similar situation. Experiments are preceded by negotiations where Kanzi and Panbanisha are politely asked if they want to work, and usually there are long discussions about what they shall eat while at work, which other apes should be allowed to be in the lab area, or which activities the apes are to be granted permission to engage in later because they agree to participate in the test work.

During the tests, the apes repeatedly are reminded of rules that must be obeyed during a controlled experiment. In the TV-documentary *Kanzi I*, for example, Kanzi participates in a word comprehension task. He sits on a chair before a table on which several photos are placed. Sue stands behind Kanzi, invisible to him, and asks him to see if he can find 'the picture of mushrooms' or 'the picture of Panbanisha' or 'the picture of keys', etc. In one instance, when Kanzi turns round to give Sue a photo, he remains in a position where she is visible to him and might unintentionally cue him. She therefore says, 'Can you turn back around': Kanzi immediately turns towards the table and awaits the next task. It happens so naturally that one scarcely notices it.

This incident allows us to glimpse Kanzi's and Sue's more familiar relation at home. Kanzi is not a laboratory animal who was specifically trained by an experimenter to hand over photos in response to hearing 'keys', 'Panbanisha' or 'mushrooms' being pronounced in a standardized fashion. The skills he draws on when he takes the test developed outside the test activity, in forms of life where Sue functioned more like a parent than like an experimenter, and where Kanzi could beg, KEY, KEY, in order to be able to open a door and enter the room where his bonobo mother, Matata, is.[28] When Sue, during the test, says, 'Can you turn back around,' as an adult can instruct a child when they visit the doctor, her speech and Kanzi's response are not properly part of the formal test. These conversations belong to the informal home framework in which Kanzi is brought into the test situation and functions there.

Home/lab duality creates the conditions for a valid comparison of ape and child, for language begins in the circumstances of home.

10. Can enculturated apes point?

What implications does what I have said here have for comparative psychology and the comparison of human and animal? I said that comparative psychology aspires to be science. However, its dependence on metaphysical attitudes, its tendency to compare

---

[27] See William M. Fields, 'Ethnographic Kanzi versus Empirical Kanzi: On the Distinction Between "Home" and "Laboratory" in the Lives of Enculturated Apes,' *Revista di Analisi del Testo Filosofico, Letterario e Figurativeo*, 8 (2007), 171–207.

[28] See the documentary, *Kanzi I*; see also Segerdahl et al., *Kanzi's Primal Language*, 59, for a description of the filmed event that I have in mind.

humans with 'all other animals on the planet', its speculative tendency to invent all-too-explanatory 'first mental causes', seem to threaten these scientific ambitions.

Here is an illustration of this threat. An influential comparative psychologist, Michael Tomasello, recently suggested that a productive way of uncovering the first mental difference that causes all other differences between humans and animals is to focus on one of the simplest of the derived differences, namely, the fact that only humans use pointing in order to convey information, and apes do not.[29] What is it about the human mind that causes humans, and only humans, to point communicatively, Tomasello asks.

But stop! How do we know that only humans point and chimpanzees do not? Well, here is the cited experiment: captive chimpanzees that have been reared by other captive chimpanzees in a lab environment are placed so that they can see how food is being hidden in one of three buckets, but they cannot see which bucket. Then, an experimenter enters and performs standardized pointing to the bucket with food. It turns out that the chimpanzees do not get the communicative meaning of this human hand movement, but perform at chance level when they try to obtain the food. When two-year-old children are subjected to the same test, however, they pass the test.

Given what we have begun to understand about enculturation and the primacy of home over lab, we need to ask: does the cited experiment compare apes and humans on the same task? Does it not compare something else too? Does it not compare two types of rearing?

The captive chimpanzees were not reared at home to collaborate with humans, and to use pointing on a daily basis to help one another in joint activities. The children, on the other hand, already point at home when they are taken into the lab and are tested there, and they are accustomed to being helped by communicative gestures like pointing. Does not the obvious asymmetry in rearing – the asymmetry in home circumstances – make the experimental comparison problematic? The fact that this concern is not voiced and studied in comparative psychology makes me think of the following statement by Wittgenstein:

> The confusion and barrenness of psychology is not to be explained by calling it a "young science"; its state is not comparable with that of physics, for instance, in its beginnings. . . . For in psychology there are experimental methods and *conceptual confusion*.[30]

The reason why comparative psychologists do not test what ought to be obvious to test, namely, whether human-enculturated apes point, is not that the field is a 'young science'. It is rather that seeing the significance of rearing, and thereby the significance of enculturated apes – the priority of home over lab – is connected with philosophical difficulties comparable to those with which Nietzsche and Wittgenstein struggled.

---

[29] Michael Tomasello, 'Why Don't Apes Point?' in *Roots of Human Sociality: Culture, Cognition and Interaction*, ed. N. J. Enfield and Stephen C. Levinson (Oxford: Berg, 2006).

[30] Wittgenstein, *Philosophical Investigations*, p. 232.

Comparative psychology faces not only scientific difficulties, but also philosophical ones. The difficulty of achieving home/lab duality, and of seeing its fundamental significance, is simultaneously the difficulty of achieving a combined effort of philosophical and experimental work in this field. In order for comparative psychology to live up to its scientific ambitions, it needs to become 'philosophized' as thoroughly as Kanzi became 'humanized'.

Do relevantly reared apes point? Do they understand a pointing human who says, 'Look, the monster is up on the roof'? They do point and respond appropriately to human pointing, but is it *real* pointing or only *apparent* pointing (without proper mental cause)? The question of enculturated bonobo pointing is explored in a recent publication by Janni Pedersen, a Danish researcher who has been working with the bonobos.[31] Here, I will simply quote one indicative example of what 'appears' to be Kanzi pointing, taken from the big language comprehension test in a monograph by Savage-Rumbaugh and fellow researchers:

[Sentence 578] *Show me the can opener. Where is the can opener?* (Kanzi points to the television. The can opener is on the television.)[32]

In conclusion: rather than 'first mental causes', *rearing* (or enculturation) is the beginning that comparative psychologists need to acknowledge *as* the beginning. Rearing is where our philosophical assumptions about human and nonhuman things are most dramatically challenged. To paraphrase Nietzsche: where *you* look for ideal mental causes in the lab, *I* see varieties of rearing in the home of culture. Ape language research experimentally disrupts metaphysical habits of thought. What hitherto was considered a distraction is what is most important.

---

[31] Pedersen et al., 'Why Apes Point: Indexical Pointing in Spontaneous Conversation of Language-Competent Pan/Homo Bonobos,' in *Primatology: Theories, Methods and Research.*
[32] Savage-Rumbaugh et al., *Language Comprehension in Ape and Child*, 192.

2

Ethics and Language: What we Owe to Speakers

David Cockburn

1. In a passage, much quoted with approval, Bentham writes: 'The question is not, Can they *reason*? nor, Can they *talk*? but, Can they *suffer*?'[1] Bentham should not be read as suggesting that the fact that a creature can talk is never relevant to how it is to be treated. He would, presumably, have acknowledged that the possibilities of my causing another pleasure or pain are transformed by the fact that we speak a common language; and might have endorsed the widely defended view that the possession of a language creates the possibility of certain thoughts, and so of forms of pleasurable or painful emotions, that are absent without it. Bentham's point, then, will be that an ability to talk is of no ethical significance *in itself*. It is only by way of its connection with something else that a capacity for speech might have a bearing on how a creature should be treated.

That general claim is, I believe, widely shared. It comes in two broad forms. (Both of which may come with or without an accompanying thought that a capacity for speech is a clear line of demarcation between humans and animals.) We may suppose that a capacity for speech is a *reflection* of something in a creature that makes it deserving of special forms of concern or respect; or we may suppose (as I suggested with Bentham) that the possession of language *creates* certain possibilities for a creature that are relevant to our treatment of it – possibilities, for example, of certain forms of suffering.

An idea of the first form is, perhaps, clear in the following remark by Aristotle:

> Voice is a sign of painful and pleasant, which is why it belongs to the other animals as well. For their nature reaches as far as having a sense of the painful and pleasant and signaling these to each other. But speech is for revealing benefit and harm, and hence too justice and injustice. For it is a unique property of man as against the other animals that he alone has a sense of good and bad, just and unjust, and so on.[2]

---

[1] Jeremy Bentham, *An Introduction to the Principles of Morals and Legislation* (London: Athlone Press, 1970), 282. Bentham's remark does, interestingly, follow his observation that: 'a full-grown horse or dog, is beyond comparison a more rational, as well as a more conversible animal, than an infant of a day, or a week, or even a month, old'.

[2] Aristotle, *Politics* 1.2, 1253a10–18.

Similarly, though with a different emphasis, the Stoics argued that 'it is not uttered speech but internal speech by which man differs from non-rational animals; for crows and parrots and jays utter articulate sounds'.[3] As Christopher Gill remarks, the suggestion is that it is the rationality that informs internal language that is the decisive factor.[4] It is on the basis of their lack of *reason* that the Stoics denied a requirement of justice to animals.[5] Again, Descartes, while insisting that it is speech that most clearly differentiates man from animals, indicates that the significance of their lack of speech lies in what it *reveals* about non-human creatures: that they lack certain crucial forms of consciousness.[6] Mary Midgley, while bringing to the issue a very different perspective from Descartes, nevertheless shares something important with him in her response to the question: '[W]hat follows if the chimps can, in some sense, talk?' 'Why', she asks, 'does it matter so much?' She continues: 'It is not just the fact that a human being talks which gives him a claim to be treated with respect. It is what his talk shows – and he shows the same thing in other ways as well, through his actions.'[7]

Elements of the view that the possession of language does not so much *reflect* as *create* certain possibilities relevant to our treatment of a creature can, perhaps, also be found in Aristotle. Gill argues that for Aristotle 'the capacity to articulate in language is closely linked with the kinds of rationality that distinguish men from beasts' through the fact that language is a crucial means to the development of such rationality.[8] More common today is the idea that the morally significant characteristics made possible by language possession are certain forms of interests, or certain forms of suffering or pleasure. Thus, R. G. Frey argues that possessing language is a necessary condition for having beliefs, that having beliefs is a necessary condition for having desires, and that having desires is, in turn, a necessary condition for having interests in the sense that is relevant to having rights.[9] On other, more consequentialist, approaches it is suggested that possession of language is a condition that makes possible the possession of concepts: this, in turn, being a condition that makes possible certain distinctive forms of pleasure or suffering that may bear on how a creature is to be treated.

In opposition to views of both of these forms I will defend two claims. I will argue, first, that the ethical significance of speech does not lie simply in its evidential or instrumental relations to something *else*; but, rather, stands along side, for example, a capacity to suffer physical pain as an aspect of a creature's nature that is relevant to the demands it makes on us. Second, in identifying a creature's capacity for certain forms of rationality or suffering as the real locus of moral significance here, these views do,

[3] A. A. Long and D. N. Sedley, *The Hellenistic Philosophers* (Cambridge: Cambridge University Press, 1987), 53T. Cited by Christopher Gill, 'Is There a Concept of Person in Greek Philosophy?' in *Companions to Ancient Thought 2: Psychology*, ed. Stephen Everson (Cambridge: Cambridge University Press, 1991), 187.
[4] Gill, ibid.
[5] Richard Sorabji, *Animal Minds and Human Morals* (London: Duckworth, 1993), 2.
[6] See Descartes's Letter to Henry More, 5 February 1649, in *Descartes: Philosophical Letters*, ed. and trans. Anthony Kenny (Oxford: Oxford University Press, 1970), 243–5.
[7] Mary Midgley, *Beast and Man* (London: Methuen, 1978), 216, 225
[8] Gill, 'Is There a Concept of Person in Greek Philosophy?', 176–9. See Aristotle, *De Sensu* 1.437a11–17.
[9] R. G. Frey, *Interests and Rights* (Oxford: Oxford University Press, 1980), 86–100.

in an important sense, stand the situation on its head; for the demands that go with those features are, in fact, aspects of what we owe to another as a speaker. While I will say little directly on the question of language in animals, I believe that my discussion will cut against fundamental motivations of the idea, which sometimes comes with views of the form sketched above, that a capacity for speech marks a sharp line of demarcation between human beings and other creatures.

2. The suggestion that an ability to talk is of no ethical significance *in itself* has considerable plausibility given a certain familiar view of the nature of language and its place in our lives: a view according to which a language is, at its core, a system of signs that enables one individual to convey to another how things stand in some portion of the world. We might, however, turn this round – taking the *prima facie implausibility* of Bentham's suggestion as grounds for doubt about such a view of language. For it might seem clear that someone's having speech places on others a wide range of demands that are absent in relation to one who does not[10]: a demand to ask her before doing certain things to her, even where what is proposed is clearly in her own best interest; to respond to her when she addresses me; to tell her certain things, including, in some cases, things she would rather not hear; to apologize to her when I have treated her badly; to point out that what she has just said is badly confused; and so on. The obligations in these cases do not derive from ideas about possible *consequences* for her, or anyone else's, pain or pleasure. I may judge that I have no choice but to say something in the face of consequences that neither she nor I will welcome; again, I may judge that the demand on me to provide some help to her is significantly altered by her *asking* for help even though I was already fully aware of her need.[11]

I should highlight two points about my examples before moving on. First, what is at issue in them is not, at least in any straightforward sense, a matter of treating one who has language *better* than one who does not; and is, in no sense, a matter of giving greater weight to the interests of the one than to those of the other. What is at issue is not a form of *preferential* treatment, based on the supposed fact that a particular creature meets some standard, but, rather, of treatment that involves a proper acknowledgement of the kind of creature with which one is confronted. And second, in referring to 'language' I am referring to *speaking*, and understanding the speech of others: that is, to forms of interaction that occur in particular contexts between bodily creatures.

3. I will return to the idea of a connection between language possession and what we may owe to a creature. Before that, however, I want to say something about the judgement that a creature has language: about what is being claimed when it is suggested that language fundamentally marks off human beings from other creatures.

---

[10] I will focus on the case in which another places demands on *me* through speaking a language I speak and my standing in some form of proximity to her. While this is, in a sense, the most funda-mental case, a fuller treatment would require consideration of the spectrum of 'impersonal' cases.

[11] In other cases, for example when I apologize or express my gratitude, while my words may bring her pleasure or relieve pain, their doing so is *dependent on* the thought that this is what must be said – not the other way round.

We should note, first, that there is a problem with my question. For we sometimes talk of things as 'speaking' or as 'having language' in ways that, as we might put it, are not to be taken entirely 'seriously'. There are speaking clocks, computer language and body language. Before the question of whether language possession may mark off human beings from other creatures can come up for serious consideration we need a way of spelling out the particular sense of 'having a language' that is at issue. Now perhaps the most natural way to do that would be to speak in terms of a 'primary', 'serious', sense of the terms 'speech' and 'language' – namely, those that have application to human beings. The speaking clock clearly does not speak in *that* sense. But while this may be fair enough, one might, given our present purposes, be concerned that there may be some question begging here. To say that human beings have more of, in a purer form, language 'in the human sense' is, one might think, hardly to identify in illuminating terms something significant that may mark human beings off from others.

That said, it is tempting to think that there must be *something* right in that approach; and, what is more, that what is right in it may go a good way towards justifying the kind of linkage between the ideas of language and humanity that recur through the history of philosophy. Consider Wittgenstein's remark: 'only of a living human being and what resembles (behaves like) a living human being can one say: it has sensations; it sees; is blind; hears; is deaf; is conscious or unconscious'.[12] This suggestion seems to rest on the idea that a human being is our 'paradigm' of a creature to which these states can be ascribed. But are there any reasons to accept this? A child may grow up in an environment in which ascriptions of 'pain' or 'fear' to non-human creatures – for example, to the family dog – are as basic as such ascriptions to human beings. Such a protest would not, however, have the same force in connection with the thesis that human beings are the paradigm of creatures to which *language* can be ascribed. For it is clear, I take it,[13] that it is in connection with *human beings* telling each other things, asking questions, saying 'Hello', and so on, that a child first grasps the range of notions that we may cover with the umbrella terms 'speech' and 'language'. It might, then, seem clear that there is something to be said for the claim that: to think of some beings as having 'language' is to think of them as interacting with each other in ways importantly analogous to those in which *we* (human beings) interact with each other in speech. For, we might reasonably argue, if there were no substantial analogies it is unclear what would be the significance of identifying as 'language' anything that takes place in the life of a non-human creature. Perhaps, then, we can say: 'Only of a living human being, and what resembles (behaves like) a living human being, can one say: it has a language; it speaks.'

The proposal might be taken in two rather different ways. I will consider these in turn.

Consider, first, the suggestion that 'behaving like a living human being' is a condition that something must satisfy if we are to be able to identify it as speaking: in the sense

---

[12] Ludwig Wittgenstein, *Philosophical Investigations*, trans. G. E. M. Anscombe, 2nd edn (Oxford: Blackwell, 1958), §281.

[13] My slight hesitation here relates to the role that stories featuring talking animals may play in a young child's life. I will, with a not entirely easy conscience, bracket these hesitations here.

that it is only against the background of something resembling human behaviour that the sounds emitted by a creature could be language. This suggestion presupposes that we have a conception of what it is to 'behave like a living human being' that is independent of the fact that human beings *speak to each other*: as if we could take out the speaking and be left with a residue of the behaviour in which it is embedded and which is a condition of the possibility of speaking; as if we could recognize creatures – a group of human beings, of bonobo, or whatever – as 'behaving like us' independently of thinking of them as creatures that speak to each other. But if we do not think of what is going on between them as speech, how much will be left of the idea that 'in every other respect' their interactions with each other might be significantly like ours? Consider, for example, a situation in which a person responds with aggression towards someone who tells him a rather shattering home truth; or with warmth towards someone who declares her affection for him. Now we might hold on to the idea that he is responding to the other with aggression or warmth in the absence of understanding what was said. But remove the idea that *something*, and, in each case, something in a certain range, was said to him by the other and our characterization of what the person did – specifically, our characterization of him as 'responding to the other' – has to change. Further, depending on the details of the particular case, his behaviour may cease to be recognizable as characteristically human: for interactions between human beings have certain distinctive shapes, none of which, in the absence of the remark that is the ground for the response, may be manifested here. Take out the insulting words that accompanied the gracious smile and the violent reaction may cease to be humanly intelligible.

Perhaps, however, that approach rests on a failure to keep in mind that speech is itself an aspect of human behaviour. Turning, then, to a second way in which we could take the Wittgenstein-inspired thought, we might hold that: we are to settle whether something in the life of others is speech by seeing how closely *it itself* resembles that which in our lives we call 'speech'.

At what level are we to seek such resemblances? Is it like this: we are to find patterns in the sounds they produce – sounds characterized in 'purely acoustic' terms – that have sufficient similarity to certain patterns in *our* lives for us to be justified in describing them in the same terms: that is, as 'language'? We may think here of a creature producing a distinctive pattern of sounds whenever it is confronted with a rabbit, another when it is confronted with something blue, and so on. But since nothing along these lines[14] bears even a distant resemblance to that which, in our lives, we identify as 'speech' we need not dwell on this proposal. If we are to identify patterns of sounds in the lives of some newly encountered creatures (human or other) as bearing a significant likeness to the patterns of sounds in *our* lives that we identify as 'speech' we will need to consider the place that such sounds have in our, and their, interactions with each other: the ways, for example, in which someone's words may be a response to what another has said or may be taken up in conversation by another. But a similarity of *that* form to the human

---

[14]   Assuming, what I doubt, that we have some real grasp on the picture being offered here.

paradigm is one to be identified only *in the light of* the thought that they are speaking. (Perhaps, indeed, only in the light of some conception of what each is saying.)

If something's likeness to what we do is to underpin the placing-with-what-*we*-do that we mark in classifying what is going on between them as 'language' it must be a likeness that can be identified independently of the idea that they are speaking. The likeness must, it seems, be at the level of 'patterns in acoustic blasts' rather than of 'things said'.[15] There is, however, no reason to suppose that at *that* level of description there are any relevant patterns to be found. Further, whatever patterns there may be at that level – the level of sound qualities measured in purely mechanical terms – most of us are quite unaware of them; and, with that, there are no grounds for the suggestion that it is in the light of any such patterns that something is to be classified as 'language'.[16]

In identifying a certain group as 'speaking' we place what they do with what we do. This will involve finding similarities between what they do and what we do: it will involve, for example, finding in their lives things such as asking questions, giving reasons for what they have said, challenging the truth of what another has said, and so on. This, however, is not to say that a creature's behaving like us is *a condition on which depends* the possibility of identifying something in its life as 'speech'. *That* idea involves a failure to recognize just how fundamental speech is to our understanding of what it is to 'behave like a human being'. There is little space for the idea that we might identify others as interacting with each other in ways importantly analogous to those in which *we* interact with each other independently of the thought that they are speaking. Identifying a likeness to the 'paradigm' presupposes, and so cannot underpin, the identification of them as speaking. 'Behaving like a human being' does not, then, state a condition that something must satisfy if we are to identify it as a speaker.

4. My question has been: what are we wondering when we wonder of some creature or group whether it is (or they are) speaking – whether they have language? We may take that question to be a request for a statement of conditions that something must satisfy in order to be a language. The aim of this section is to suggest a different way of taking, and so of answering, the question.

The plausible thought, considered in the previous section, that human language is necessarily our paradigm of language may slide into the (perhaps less tempting) thought that English is necessarily the native English speaker's paradigm of a language. We see

---

[15] While it may well be that I am working with a sharper dichotomy than is really justified this does not, I *think,* affect my central point here.

[16] See the following observation made in the context of the study of language in the bonobo: 'It is because we hear what Kanzi says to us in our daily interactions with him that we can begin to trace analogies to spoken English. The subtle distinctions in the sounds he produces, which correspond to distinctions in spoken English, are undetectable unless you already understand what he says to you. To experience the relevant analogies, you must already understand his speech. . . . So, the reason we understand the bonobos is not that we detect acoustic parallels to spoken English. It is rather because we understand what the bonobos say to us that we can trace analogies to spoken English' (Pär Segerdahl et al., *Kanzi's Primal Language* (Basingstoke: Palgrave, 2005), 64–5, 206). I should note here a substantial debt to this highly stimulating and perceptive study.

a variation on this in Davidson's suggestion that 'translatability into our language' is a condition for identifying something as a language at all.[17] That something is going wrong there should be clear from the fact that I might recognize a group as speaking by finding that, perhaps with much attention and patience over an extended period, I can learn to speak with them. Thus, one might find one's feet in a foreign linguistic community independently of considering how far what they say might be rendered in English – or how like what is going on between them is to what goes on between English speakers.

One's relation to another's words is very often not that of observer of what another does, but of participant in conversation with the other. This case is sufficiently widespread and fundamental to raise a doubt about the appropriateness of giving *the* central place in philosophical discussion of language to the perspective of the disinterested classifier reflected in the question: 'What conditions must something satisfy in order to be a language?' Perhaps, then, we will do well to shift our focus from the theoretical, third person, stance in which an individual identifies a phenomenon with which she is confronted as 'language', to the perspective of the active participant whose identification is seen in her engagement with another's speech. (A fuller justification for adopting this approach must lie in its results.)

It is in this spirit that we may ask: what is it seriously to think of – to acknowledge in practice – an individual or group as having a language, as speaking? In response to this question we might start with the following (again Wittgenstein-inspired) formulation: taking up another's words in my relations to her, or struggling to find the sense in what she is saying, are forms of the conviction that someone else is speaking to me.[18] Instances of what I mean by 'taking up another's words' would be: responding to a greeting or request, asking whether she really means what she appeared to say, pointing out an obvious confusion in what she said, connecting things she says now with what she said earlier, holding her to a promise, and so on. It is often in responses such as these that, in practice, we see most directly that someone takes another to have said something; and, at a more general level, that we see most directly that someone takes another to be a speaker. The point of the words 'most directly' here is to suggest that such responses should not be construed as the public manifestations of an underlying state of passive recognition. In standard, central, cases my responding to another in these ways is not a *consequence* of the thought that she has just said some particular thing.

It is tempting to add: to take her to have said a particular thing simply *is* to respond to her in this way. For example, in a particular case, my taking her to have asked me to pass the salt simply is to pass the salt in response to what she said. But claims, in philosophy, of the form 'This *simply is* that' should be handled with care. Thus, I have, it might plausibly be argued, not recognized this as the request that it is if a number of other conditions are not met: for example, if I would not acknowledge that I will have failed her if I do not pass the salt, if I would just as readily have passed the pepper had it been closer to hand, and perhaps much more of the same kind. Now we should not

---

[17]   See Donald Davidson, 'On the Very Idea of a Conceptual Scheme,' in his *Inquiries into Truth and Interpretation* (New York: Oxford University Press, 1984), 183–98.
[18]   See Wittgenstein, *Philosophical Investigations*, §287.

expect a definitive answer to the question: just what are the required conditions; just *how much* more of this kind is required for this to be a genuine case of taking another to have asked me to pass the salt? We are not, after all, *wrong* to say of the dog that it takes its master to have asked him to fetch his slippers; and there need be no definitive line that the child must cross for us to be able accurately to say of her that she now understands simple requests, promises, and so on. The point is simply that it is only within a wider framework of relations to another that my immediate response on this occasion has the significance that it does.

Any remotely adequate characterization of the relevant 'wider framework' would, I suspect, require a detailed consideration of what Wittgenstein gestures to with the term 'fine shades of behaviour'.[19] It would also require reference to the individual's acknowledgement of *demands* on her that arise from the other's words. Thus, I listed some responses in which we may see that someone takes another to have said something: responding to a greeting or request, asking whether she really means what she appeared to say, and so on. I now want to add: central to what marks off such a response in a particular case as a manifestation of understanding, as opposed, say, to a mere conditioned reflex, is an acknowledgement that there is a *demand* on me to respond in an appropriate way: an acknowledgement that I will have failed in what I owe to the other if I do not respond to her greeting, point out (or, in a different case, attempt to cover over) the apparent absurdity of his remark, and so on.

Consider another kind of case. While I am in no doubt that this man is speaking to me, and while I can make out the English sentences he is producing, they do not hang together in a way of which I can make much, or anything. Here, the demands in play are of a different form from those stressed so far. If I think of him as having said something I may attempt to find the sense in it; and, depending on the case, we may judge that I have an obligation to do so. Now, we may picture such situations in this way: trying to find the sense in what he is saying is trying to identify something that lies behind the words – 'in his mind' – which, if only I could see it, would cut out the need for the work. A need to *assume* that there is some sense in what the other is saying – a need for 'charity' – along with a readiness to *work* in conversation with him to draw out a sense, are only required because of an epistemological obstacle: an obstacle that would be removed if I could stand to that sense as the speaker himself does. Such images, involving the Cartesian idea of the sense of his words as something that is immediately and incorrigibly available to the speaker, are, perhaps, now widely recognized to be inadequate. It is generally agreed that nothing 'in the speaker's mind' guarantees that there is *anything* to be found, or fixes *what* will be found, by such work. The other side of that point is this: the thought that, despite the apparent chaos in his words, he really is saying something is not one that *justifies* the charity and the struggle. The attitude towards the other that we see in this kind of case is itself partly constitutive of the thought that he really is saying something.[20]

[19]   Ibid., pp. 203–4, 207.
[20]   This is not to say that the charity and struggle can have no grounds. My point is that the grounds do not include the thought that he really is saying something.

One of my aims in this essay is to clarify the relation between the idea that another makes certain demands on us and the idea that she speaks. We may be inclined to think of the relation in this way: an acknowledgement of the demands is grounded in the thought that she is speaking; that another is a speaker is something to be established independently of our taking any particular attitude towards her – with one's sense of how this creature is to be treated disconnected – and so can justify our taking the relevant attitude. If we think of the matter in that way, we will be left with the question of *how* the fact that another is a speaker would justify the attitude; and, puzzled by that question, may find ourselves drawn into pictures in which the significance for my treatment of her of the fact that another is a speaker is purely evidential or instrumental. That aside, it is, I hope, now clear that there is another way in which we might understand this relation. Perhaps it would be closer to the mark to say: in standard, central, cases my idea that another makes on me the kinds of demand that I mentioned earlier – a demand to ask her before doing certain things to her, to respond to her when she addresses me, to tell her certain things, and so on – is not a *consequence* of my taking her to be one who speaks. It is, rather, part of grasping what language is, and, with that, the distinction between those who speak and those who do not, that one accepts that the first make on me demands of a kind that the second do not.[21]

I have suggested that that might be 'closer to the mark'. How is it to be shown that it is in fact so: that there is no recognition that a creature speaks in the absence of an acceptance of such demands? In a sense it *isn't* to be shown. It is, after all, not *wrong* to say that some clocks speak. All that we can do is to think through carefully what will be left if such an acceptance is absent (as I will not try to do here), and ask ourselves whether that adequately captures what we have in mind when we think of a capacity for speech as a hugely significant aspect of human life. And people with different theoretical interests may answer that question in different ways.

5. At the start of this essay I spoke of views of this form: that a creature has language is, in one way or another, grounds for the judgement that it has something *else* – a capacity, perhaps, for certain forms of consciousness or of reasoning – this something else being, in turn, a ground of certain obligations that we have towards it. I have been suggesting that, as so often in philosophy, the search for *grounds* for certain features of our lives leads to serious distortion of those features. In opposition to views of the forms outlined, I have argued, first, that a creature's capacity for speech is, in itself, of significance in the demands that it makes on us; and, second, that the idea that a creature makes such demands on us is not *dependent on* the thought that she speaks: it is itself an aspect of taking her to speak. Taking another to have language itself involves a certain ethical relation to her.[22]

---

[21] This remark gives central place to the case in which *I* am addressed. A more rounded picture would speak also of the recognition of demands on *others*.

[22] What I spoke of as the 'Cartesian' idea may, then, be, in part, an instance of what Charles Taylor speaks of as 'The transformation of norm into theory that is so typical of modern culture' (Charles Taylor, 'Theories of Meaning,' in his *Human Agency and Language* (Cambridge: Cambridge University Press, 1985), 291).

I want now to indicate a sense in which the pictures I have rejected get things back to front. In the place that they give to reason or to particular forms of consciousness they reveal a failure to appreciate the *pervasiveness* of ways in which the possibility of interaction through speech is implicated in our relations with others.

Consider, first, reason. I will focus on the special sense in which one who has language may have reasons for what she does, thinks or feels. We can, of course, attribute to a creature that does not have language reasons for what it does; but, it is generally supposed, there is a sense of 'reason' that is (a) intimately connected with language possession, and (b) may have a bearing on what we owe to the creature. Now we may picture language possession as something that *reveals*, because it is only made possible by, a special form of rationality; or we may picture it as a condition that *makes possible* this form of rationality. But we will have a better view of the way in which a creature's possession of reason may bear on what we owe to it if we view things in neither of those ways. For one of the most immediate and obvious ways in which it does so is that it will sometimes place on us an obligation to *ask* the other for her reasons, to *offer* her reasons and to *reason with her*. What is at issue here is an obligation to relate to her *in language* in certain ways. The range of demands that come into play does not flow from the fact that her having language is indicative of something *else*: this something else – 'a capacity to reason' – being the real locus of the demands. To accept her as a creature of that kind – a kind that has reason in this sense – just *is* to accept that she is of the kind to be, for example, reasoned with.[23] We can add: the acceptance of an obligation does not *flow from*, is not *justified by*, my recognition of the fact that she is a speaker. It is an *aspect* of it.

What of the relation between language and a capacity for particular forms of pain and pleasure that may bear on our treatment of a creature: fears about, or pleasant anticipations of, the relatively distant future; regrets about, or comforting memories of, the relatively remote past; and so on? We may think of the matter in this way:

> The possession of language is a condition that makes possible the possession of concepts, and in particular of certain temporal concepts. This in turn is a condition that makes possible a creature's reaching out in thought to events in the relatively remote past or future: making them available as objects of concern, and so as possible sources of pleasure or pain. Of course, the pleasures or pains of, for example, anticipation are qualitatively different in significant ways from the pleasures or pains of warm baths and crushed toes. But it is what they share that makes them both proper objects of concern: makes them both states that should enter into our understanding of what we owe to another. Thus, the significance of language for our treatment of a creature can be acknowledged within a basically utilitarian framework. That a creature can talk does not, in itself, have a bearing on what we may owe to it: it does so only through what its possession of language makes possible for it.

---

[23] In a more ontological mode we might say: to *be* a creature of that kind – a kind that has reason in this sense – just is to be of the kind that is, for example, to be reasoned with.

But what *do* grief and physical pain share that makes them both forms of suffering – both proper objects of concern by others? If we are tempted to think of this in terms of some common introspectable quality, a moment's reflection should leave us puzzled. Following Wittgenstein,[24] we may ask: suppose one minute of the state of grief 'could be isolated, cut out of its context; would what happened in it then not be' grief? We might ask further: would what happened in it then not be suffering? And if we are inclined to answer 'It would still be suffering', we can ask: is it a perceived qualitative likeness to what I feel when I stub my toe that licenses that answer?

We speak of physical pain, grief and fearful anticipation in terms that, up to a point, run parallel: characterizing them all, for example, as forms of 'suffering'. Our use of the same word – 'suffering' – is not simply an instance of linguistic ambiguity. This is so, not in virtue of some 'perceived common quality in what we are speaking of', but in virtue of the fact that that common use is an aspect of a wider network of ways in which we relate to these varied phenomena – ways that, up to a point, run parallel. It is *this* that is marked in our language by our use of the word 'suffering' in relation to all these conditions. We might summarize something that runs through this wider network in the following, slightly shaky, terms: both grief and physical pain are typically states we want to avoid, both for ourselves and for those we care about, and people in those states are, typically, to be pitied and comforted.

The summary is shaky in that avoidance, pity and comfort themselves take different forms. In particular, in relation to one with whom I can speak, forms of acknowledgement of her suffering are available that would not otherwise be: I may reassure her that a physical pain will be over soon. Further, with the possibility of new forms of acknowledgement comes the possibility of the acknowledgement of states for which there was no place before. Thus, my recognition that another is feeling shame about something she has done may be expressed in attempts to convince her that she has nothing to be ashamed of; or in attempts to show her what she can do now in view of her admittedly shameful treatment of another. I may, of course, recognize that another is feeling shame without speaking to her in such ways; but if I do not appreciate that, depending on the details of the case and my relation to her, I, and others, may owe it to her to speak in ways such as these, there may be little room for the idea that I recognize it to be *shame* that she is feeling (as opposed, perhaps, to simple regret.)

We may say that it is a 'logical', or 'conceptual', truth that only a creature with language can feel certain forms of shame or the consolation that comes with knowing that the pain will soon be over. But what is right in that may be better expressed in this way: one who seriously judges another to be in such a state is one who judges that she is to be *talked to* in certain ways – in central cases, is one who accepts that he owes it to her to talk to her in certain ways. There is, then, no place in the life of one who lacks language for such seriously judging; and no place for such serious judging when the object of my concern is a creature that lacks language.

I opened my paper with Bentham's remark: 'The question is not, Can they *reason*? nor, Can they *talk*? but, Can they *suffer*?' My suggestion now is that in his proposal that this is *the* question Bentham's appeal to the contrasts between 'reason' and 'suffering' on

---

[24] Wittgenstein, *Philosophical Investigations*, §584.

the one hand, and 'talking' on the other, must be questioned. Thus, while there are no grounds for the claim that only a creature that can talk can suffer, in relation to certain forms of suffering the fact that a creature can talk in certain ways, and, with that, that there is a demand on us to talk with him in certain ways, is not to be contrasted with the fact that he can suffer in these ways. A recognition that I owe it to the other to relate to him in language in certain ways is not *grounded* in a recognition that he has a capacity for this form of suffering.

6. I have said: taking another to have language itself involves a certain ethical relation to her. That is to say, what is at issue when we wonder whether a creature speaks is a question about how we are to relate to it. There is no establishing that another is a speaker independently of taking a certain moral stance towards her – the first being a ground for the second.[25] This is not to suggest that the stance may be adopted at will, independently of the nature of its object. It is in virtue of the kind of creature this is that it makes on me the demands that it does; and, in cases of uncertainty, what kind of creature this is is to be settled by close attention to *it*. It does not follow that there is any identifying of the relevant features of the creature's nature independently of an acknowledgement of the demands the creature places on me: a demand, for example, to greet her, address her by her name or apologize for treading on her.

Where does this leave the possibility of language in non-human animals? I will close with some brief remarks on this.

First, while my emphasis has been on the notions of *speaking* and being a speaker, I have moved fairly freely between those locutions and talk of a creature's 'having language'. There are dangers in that. One is that one may readily bypass the possibility of a creature's *understanding* speech while not itself having a capacity for speech. Attributions to a dog, or young child, of understanding things we say to them may be quite rich despite the relative, or total, absence of any 'speech' on their part. At this point, there is very little to be gained by pressing the question 'But do they have language?' Indeed, an insistence on the question may well be the expression of philosophical confusion.

Another danger here lies in failing to acknowledge a distinction between speaking and having *a language* – in the sense in which to speak English, French or Urdu is to have a language. I will not attempt to say what that sense is. It is, I take it, what Rush Rhees has in mind when, in his discussion of Wittgenstein's builders, he remarks that language is something that can have a literature.[26] It is, one might add, something that can have a dictionary – where a 'dictionary' is not *simply* a record of how people speak. It may be suggested, plausibly I would think, that any analogue of this – of *a language* – in the lives of non-human creatures is distant. That, however, leaves completely open the question whether we may relate to them, and they to each other, in ways that can, in a rich sense, be spoken of as 'speech'.

---

[25] Compare the following remark: 'We cannot point and say, "This *thing* (whatever concepts it may fall under) is at any rate capable of suffering, so we ought not to make it suffer" ' (Cora Diamond, 'Eating Meat and Eating People,' in her *The Realistic Spirit* (Cambridge, MA: MIT Press, 1991), 325). I owe a great debt to this marvellous paper.

[26] Rush Rhees, *Wittgenstein and the Possibility of Discourse* (Cambridge: Cambridge University Press, 1998), 192.

My emphasis has been exclusively on the ethical relations involved in being able to speak with another. This leaves open questions relating to the possibility of recognizing that a creature has language in contexts in which, not *sharing* a language with her, I cannot speak with her. There is a sense in which this is the situation in which we all stand to most other human beings. This sense is, I think, very different from that involved in theoretical speculations about the possibility that some animals may communicate with each other in language: language that we do not, perhaps could not, understand. The following remark in the Kanzi book is directly relevant to such speculation:

> Language is not an inanimate object. It is unclear what it would mean to sit undisturbed outside an enclosure and decide objectively whether that object is inside. . . . Morally, then, there is no such thing as a neutral observer of the apes' language.[27]

While this seems to me both fairly startling and correct, I will settle here for a more modest claim. If one supposes that having language is, directly or indirectly, a mark of a creature's meeting some standard that confers on it a right to forms of preferential treatment by us, the possibility that some animals may, unknown to us, have language will be of substantial ethical concern. If, however, one believes, as I have suggested, that the ethical issues involved here are not of *that* form, little may be left of the moral urgency of the speculations. I am tempted to say that little, too, may be left of the speculations themselves. But that goes beyond anything I have given significant grounds for.

Finally, I said that I hoped that my discussion would cut against the idea that a capacity for speech marks a sharp line of demarcation between humans and other creatures. So long as we suppose that a creature's 'having speech', or 'having language', is a condition that *grounds* certain obligations that we have towards it, we may be inclined to think of it as something whose presence is to be investigated by the experts while they 'sit undisturbed outside an enclosure and decide objectively whether that object is inside'. If we abandon that picture, the question 'Do they have a capacity for speech?' may appear in a very different light. Our question becomes one that calls for an exploration of the ways in which we do, or, with the right kind of work, may come to, relate to the creatures in question. While my personal experience of animals is too limited for to me to be able to say anything useful about how far such ways of relating might be possible, and demanded, in our relations with certain non-human creatures, I am persuaded by conversations with, and work by, people with relevant experience of dogs and bonobo that it is so to a much greater degree than I would have expected.[28]

[27] Segerdahl et al., *Kanzi's Primal Language*, 89–90.
[28] I am grateful to a number of people for very helpful comments on earlier drafts of this essay. I would like, in particular, to thank Patrick Cockburn, Andrew Gleeson, Anniken Greve, Lars Hertzberg, Maureen Meehan, David Robjant, Lynne Sharpe, Lloyd Strickland and participants at the Nordic Wittgenstein Society conference on *Language, Ethics, and Animal Life* held in Uppsala in March 2010.

3

The Difficulty of Language: Wittgenstein on Animals and Humans

Nancy E. Baker

The logic of language is immeasurably more complicated than it looks.[1]

There have been several ways of radically differentiating human beings from animals in the history of philosophy, one of which is to use language as the differentia. Typically, these attempts have been for the purpose of defining the essence of 'human being' by using animals as an oppositional foil. Presupposed in this enterprise has been a hierarchy in which human beings are ranked as 'higher' and their experiences more valuable. Although Wittgenstein has been assumed to be 'anthropocentric' in this sense because of his focus on language,[2] he in fact has much to offer those who are calling into question the metaphysics of the animal–human divide.[3]

I have borrowed for the title of this essay from Cora Diamond's 'The Difficulty of Reality and the Difficulty of Philosophy'.[4] The 'reality' in her paper is the horror of what we do to animals, particularly in the food industry, as experienced by J. M. Coetzee's Elizabeth Costello in his novel by that name. The 'difficulty' is 'the mind's not being able to encompass something which it encounters'.[5] As a result Elizabeth Costello 'does not engage with others in argument, in the sense in which philosophers do'.[6] Diamond elaborates on this as follows: '... "debate" as we understand it may have built into it a distancing of ourselves from our sense of our own bodily life and our capacity to respond to and to imagine the bodily life of others'.[7] Although the difficulty and outcome are very different, I see an analogy here with what Wittgenstein shows us about language

[1] Ludwig Wittgenstein, *Last Writings on the Philosophy of Psychology*, Vol. II (Oxford: Blackwell, 1992), 44e. Hereafter abbreviated to *Last Writings*, II.
[2] For an interpretation of Wittgenstein as 'anthropocentric' see Michael P. T. Leahy, *Against Liberation* (London: Routledge, 1991). For an excellent criticism of this interpretation of Wittgenstein, see Nigel Pleasants, 'Nonsense on Stilts? Wittgenstein, Ethics, and the Lives of Animals', *Inquiry*, 49 (2006), 314–36.
[3] For an account of this questioning in Heidegger, Levinas, Agamben and Derrida see Matthew Calarco, *Zoographies* (New York: Columbia University Press, 2008).
[4] In *Philosophy and Animal Life* (New York: Columbia University Press, 2008).
[5] Diamond, 'The Difficulty of Reality and the Difficulty of Philosophy', 44.
[6] Ibid., 52.
[7] Ibid., 53.

and philosophy. There are actually several difficulties. First, for him there is a difference between philosophy in the traditional sense of theory, generalization, and argument, on the one hand, and, on the other hand, the philosophizing he calls 'investigation'. The latter begins not with a distancing, disembodied position to be defended, or the continuation of a debate, but very personally with confusion and not knowing one's way about.[8] Another difficulty is what is required for the way out of that confusion: We need to attend to how language is actually used, a reality the mind is not able to encompass with its abstractions and generalizations, namely, at a distance: 'Nothing is more difficult than facing concepts *without prejudice*. (And that is the principal difficulty of philosophy.)'[9]

The investigations Wittgenstein undertakes are 'conceptual investigations'[10] and require 'quiet weighing of the linguistic facts'.[11] As he reminds us over and over, we are tempted to evade both the difficulty of confusion and the difficulty of immersing ourselves in the details of our language. I would add another difficulty here – that of reading Wittgenstein's actual words, as opposed to using isolated quotations, or even just a reputation, for one side or another of a debate, in this case a debate about animals.

Wittgenstein mentions animals of various kinds throughout his later work. His use of animals, as well as children, has everything to do with his overall project of clearing up conceptual confusions about the 'inner' and other minds, about what speaking a language is, about criteria for the application of our mental concepts, and even about foundations, whether of mathematics, language or knowledge. Not only is there no opposition in his work between humans and animals, but, in keeping with his anti-essentialism, animals are not lumped together in one category, a requirement of oppositional thinking. In fact, as Ray Monk reports in his biography, Wittgenstein derived great pleasure from the earth's immense variety of fauna and flora, and even thought that Darwin's theory 'hasn't the necessary multiplicity'.[12] The theme of multiplicity appears throughout his later work and is expressed in his desire to use the Earl of Kent's phrase from *King Lear*: 'I'll teach you differences' as a motto for *Philosophical Investigations*.[13] Significantly, the concept of 'language' itself has no essence, something not noticed by those who take Wittgenstein to be using language as the differentia to define the essence of humans.

Wittgenstein had a strong sense of development, both phylogenetic and ontogenetic, and, again, in keeping with his anti-essentialism said there is 'no sharp boundary'

[8] See Ludwig Wittgenstein, *Philosophical Investigations*, revised 4th edn (Malden, MA: Wiley-Blackwell, 2009), §123. All quotations from *Philosophical Investigations* will be from this edition unless otherwise stated.
[9] Ludwig Wittgenstein, *Remarks on the Philosophy of Psychology*, II (Chicago, IL: University of Chicago Press, 1980), §87. Cf. Ludwig Wittgenstein, *Last Writings on the Philosophy of Psychology*, Vol. I (Chicago, IL: University of Chicago Press, 1982), §12. Hereafter abbreviated to *Last Writings*, I.
[10] Ludwig Wittgenstein, *Zettel* (Berkeley, CA: University of California Press, 1967), §458.
[11] Ibid., §447. See also Wittgenstein, *Philosophical Investigations*, §§51–2.
[12] Ray Monk, *Ludwig Wittgenstein: The Duty of Genius* (New York: The Free Press, 1990), 537.
[13] See M. O'C. Drury, 'Conversations with Wittgenstein,' in *Recollections of Wittgenstein*, ed. Rush Rhees (Oxford: Oxford University Press, 1984), 157.

to be drawn between levels.[14] Most importantly, he reminds us that even our most sophisticated human cognitive abilities, including that of speaking a language, are themselves forms of *behaviour*. This is to say that the criteria for the application of our mental concepts involve the *body*. It would, then, seem that it is not just 'debate' that distances us 'from our sense of our own bodily life and our capacity to respond to and to imagine the bodily life of others', as Diamond puts it. It is also the Cartesian radical separation of mind and body at the heart of certain anthropocentric criticisms of 'anthropomorphism', namely, the presumed projection onto animals of characteristics assumed to be exclusively human. What is actually at issue here, as Wittgenstein shows us, is seeing what the criteria are for the application of our mental concepts and how that sheds light on when and why it is appropriate to speak of animals in apparently human terms, whether in ordinary life, veterinary medicine or the field of animal behaviour. Needless to say, a great deal is now known about animal behaviour that was not known when Wittgenstein was alive. This should make little difference, since Wittgenstein's investigations are conceptual and, as we shall see, what he has to teach us applies as well to new empirical data.

In light of some of the above mentioned issues, I would like to look at what Wittgenstein actually says about our language concerning animals, and have, for this reason, quoted extensively from the texts. This will enable us to see why for him, contrary to a deep intellectual thread in Western thought, there is *no essential difference in kind* between animals and humans, and why, surprisingly, despite many obvious differences between us, it is not quite right to say that animals *lack* anything that human beings have.

1. Behaviour

Forms of behaviour may be incommensurable.[15]

What makes it possible to use some of the same language for human and animal behaviour in a meaningful way is, to use Wittgenstein's word, the *resemblance* between us, not only in the construction of our bodies but also in our behaviour. Even when we don't look alike, we do all interact with our environment and with other creatures. In other words, we *behave*: 'only of a living human being and what resembles (behaves like) a living human being can one say: it has sensations; it sees; is blind; hears; is deaf; is conscious or unconscious'.[16]

What exactly is meant by 'behaviour'? Wittgenstein reminds us that there are many kinds, for example, that of planets, machines, robots, automatons, amoebas, flies, cats, dogs, the feeble-minded, idiots, imagined Martians, infants, children, the deaf,

---

[14] Ludwig Wittgenstein, *Philosophical Grammar* (Berkeley, CA: University of California Press, 1974), 62.
[15] Wittgenstein, *Remarks on the Philosophy of Psychology*, I, §314.
[16] Wittgenstein, *Philosophical Investigations*, §281.

the colour-blind, imagined tribes of human beings and normal adults. In the case of animals and human beings, behaviour requires a body and one that *moves* and *changes* in certain ways.

> If you want to act like a robot – how does your behaviour deviate from our ordinary behaviour? By the fact that our ordinary movements cannot even approximately be described by means of geometrical concepts.[17]
>
> Look at a stone and imagine it having sensations. – One says to oneself: How could one so much as get the idea of ascribing a *sensation* to a *thing*? One might as well ascribe it to a number! – And now look at a wriggling fly and at once these difficulties vanish and pain seems able to get *a foothold* here, where before everything was, so to speak, too *smooth* for it.[18]

The face, too, becomes an important part of behaviour certainly in human beings and in many animals as well. In imagining a debate about whether dogs have souls, Wittgenstein suggests that one side might say, 'Just look at the face and the movements of a dog, and you'll see that it has a soul,' and then he asks what it is we see – among other things, a 'lack of stiffness'.[19]

On the other hand, if we hold an exact representation of a person's face upside down, we most likely won't be able to 'tell the *expression* of the face'.[20] That there is something more to behaviour than what can be exactly geometrically represented of the arrangement or movements of body parts is explored further in a contrast between behaviour and anatomy: 'For how could I see that this posture was hesitant before I knew that it was a posture and not the anatomy of the animal?'[21] These two aspects of both animal and human behaviour – the anatomical or geometrical, on the one hand, and the expressive, on the other – are described as two different *levels*: ' "I see that the child wants to touch the dog, but doesn't dare." How could I see that? – Is this description of what is seen on the same level as a description of moving shapes and colours?'[22] No, and of someone who fails to see in a face something beyond the geometrical we might say 'that he was blind to the *expression* of a face'.[23] As we know, this kind of 'blindness' can happen in certain forms of autism.

Context creates yet another level of behaviour. For example, we could describe or study the anatomy of a cat's running and compare it with its walking, or we could compare it with the running or walking of a dog. If, however, we add particular surroundings or context the word 'behaviour' is given a broader, more complex meaning, and the behaviour in question is, then, at a different level. The cat's running

---

17 Wittgenstein, *Remarks on the Philosophy of Psychology*, I, §324.
18 Wittgenstein, *Philosophical Investigations*, §284.
19 Wittgenstein, *Last Writings*, II, 65e. Cf. Wittgenstein, *Remarks on the Philosophy of Psychology*, II, §§615, 627.
20 Wittgenstein, *Remarks on the Philosophy of Psychology*, I, §991.
21 Wittgenstein, 'Philosophy of Psychology – A Fragment,' §225, in *Philosophical Investigations* (Malden, MA: Wiley-Blackwell, 2009), 220. Cf. Wittgenstein, *Last Writings*, I, §736.
22 Wittgenstein, *Remarks on the Philosophy of Psychology*, I, §1066; cf. §1068.
23 Wittgenstein, *Last Writings*, I, §763; cf. §770.

could be mouse-catching behaviour or chasing play behaviour or fear behaviour. The actual physical movement might be the same or different in each case, but what we call and how we describe the behaviour in this broader non-anatomical sense is determined by context or background:

> Forms of behaviour may be incommensurable. And the word "behaviour", as I am using it, is altogether misleading, for it includes in its meaning the external circumstances – of the behaviour in a narrower sense. / Can I speak of one behaviour of anger, for example, and of another of hope?[24]

To take a human example, a small child is crying. Is it in pain or emotionally distressed?[25] We need the context to decide.

2. Animal and human behaviour: Some similarities

> "So these concepts are valid only for the *total* human being?" – No, for some have their application to animals too.[26]

Wittgenstein comments further on our 'seeing' the shy behaviour of the child who wants to touch the dog but doesn't dare by again distinguishing different levels of behaviour and adding a dog:

> . . . one does say, that one sees both the dog's movement and its joy. If one shuts one's eyes one can see neither the one nor the other. But if one says of someone who could accurately reproduce the movement of the dog in some fashion in pictures, that he saw all there was to *see*, *he* would not have to recognize the dog's joy.[27]

What is to be noticed here, in addition to the fact that both the human and the animal behaviour involve more than what can be represented 'geometrically', is that we *do* speak of a dog's *joy*. When an owner comes home after an absence and her dog jumps up and down, wags its tail, licks, pants, runs around, brings toys, we say that the dog is *glad to see her*. It is unlikely that the dog would greet the vet in the same way. If the dog behaved this way towards the refrigerator we would not say it was *greeting* the refrigerator or *glad to see* the refrigerator. Again, context matters: 'If someone behaves in such-and-such a way under such-and-such circumstances, we say that he is sad. (We say it of a dog, too.)'[28]

A dog's wagging its tail is seen by us as happy not sad both because of the circumstances in which it typically occurs, but also because of the configuration of

[24] Wittgenstein, *Remarks on the Philosophy of Psychology*, I, §314; cf. §129.
[25] Cf. Wittgenstein, *Zettel*, §492.
[26] Wittgenstein, *Remarks on the Philosophy of Psychology*, II, §328.
[27] Wittgenstein, *Remarks on the Philosophy of Psychology*, I, §1070.
[28] Wittgenstein, *Zettel*, §526.

body parts and the rhythm and speed of their movement. The tail is up and wagging. Often there is rapid jumping up and down. If we see that a dog is sad or depressed, we see the tail down and slowness in its movements. We see similar behaviour in human beings, and even have in our language such expressions as 'jumping for joy' and 'feeling down'. In the case of a feeling of well-being, Wittgenstein says, 'Here there occurs to me the special expression of well-being. A cat's purr, say.'[29] A cat's purr typically is a continuous, even, slow-ish sound, as is the 'mmmmmm' sound of a person in the sun on a cool day, expressing well-being. In both cases the whole body is in one place, not running around or jumping up and down, and eyes likely are shut.

Because of the resemblance to human behaviour in similar contexts we say that certain animals can be *angry, frightened, sad, joyful, startled, mean, brave, hesitant*;[30] that a dog *feels fear, sadness, joy*, that it *suddenly notices*, that it *believes*;[31] that an ape *investigates an object*.[32] Even in 'an intelligent beast' we could see behaviour characteristic of *changing one's mind* or *considering*.[33]

How great does the behavioural resemblance have to be to apply the same concepts to animals as we apply to human beings? One way to consider similarities in behaviour of animals and humans is to look at our interactions. With cats and dogs and chimpanzees it is obvious, but the octopus, an animal that barely resembles us in bodily construction, is now discovered to behave in very recognizable ways, particularly when interacting with us. Those who have worked with them describe them as *shy, affectionate, interested, curious, having individual personalities* and *not liking certain researchers* to the extent that they consistently squirt them with a stream of salt water and thereby drive them away.[34]

3. Animal and human behaviour: Some differences

There is nothing astonishing about certain concepts' only being applicable to a being that e.g. possesses a language.[35]

The most striking difference between animals and humans is, of course, that human beings are language users. We speak and we do so in many different ways: We describe, tell jokes, ask for help, question, command, greet, pray, etc. Speaking a language is an activity, and, as is well known, Wittgenstein calls all these different linguistic

---

[29] Wittgenstein, *Remarks on the Philosophy of Psychology*, I, §122.
[30] See Wittgenstein, 'Philosophy of Psychology – A Fragment,' §§1, 225; Wittgenstein, *Last Writings*, I, §736.
[31] See Wittgenstein, *Zettel*, §§518, 526; Wittgenstein, *Philosophical Investigations*, §650; Wittgenstein, *Remarks on the Philosophy of Psychology*, I, §1070; Wittgenstein, *Last Writings*, I, §537; Wittgenstein, 'Philosophy of Psychology – A Fragment,' §1.
[32] Wittgenstein, *Remarks on the Philosophy of Psychology*, I, §347.
[33] Wittgenstein, *Remarks on the Philosophy of Psychology*, II, §6; I, §561.
[34] See, for example, Sy Montgomery, 'Deep Intellect: Inside the Mind of the Octopus,' *Orion Magazine*, November/December 2011.
[35] Wittgenstein, *Zettel*, §520.

activities 'language-games', which are described as 'language and the actions into which it is woven'.[36] They are forms of behaviour: 'For our *language-game* is a piece of behaviour'.[37]

If speaking a language is an activity, a form of behaviour, what is it that language users can *do* that non-language users cannot?

> One can imagine an animal angry, fearful, sad, joyful, startled. But hopeful? And why not?

> A dog believes his master is at the door. But can he also believe his master will come the day after tomorrow? – And *what* can he not do here? – How do I do it? – What answer am I supposed to give to this?

> Can only those hope who can talk? Only those who have mastered the use of a language. That is to say, the manifestations of hope are modifications of this complicated form of life. (If a concept points to a characteristic of human handwriting, it has no application to beings that do not write.)[38]

Near the beginning of *Philosophical Investigations* Wittgenstein says that 'to imagine a language means to imagine a form of life',[39] in this case a small and simple language consisting only of reports and orders in battle. A few paragraphs later he says, 'The word, "language-game" is used here to emphasize the fact that the *speaking* of language is part of an activity, or of a form of life'.[40]

Unlike the simple language of reports and orders, hope requires 'a complicated form of life', not yet had by the developing toddler:

> Someone says: "Man hopes." How should this phenomenon of natural history be described? – One might observe a child and wait until one day he manifests hope; and then one could say "Today he hoped for the first time". But surely that sounds queer! Although it would be quite natural to say "Today he said 'I hope' for the first time". And why queer? One does not say that a suckling hopes that . . ., but one does say it of a grown-up. – Well, bit by bit daily life becomes such that there is a place for hope in it.[41]

And, also 'a place for' *thankfulness*, something we do not attribute to a newborn.[42]

---

[36]   Wittgenstein, *Philosophical Investigations*, §7. Cf. Ludwig Wittgenstein, *On Certainty* (New York: Harper Torchbooks, 1972), §229; Wittgenstein, *Zettel*, §173.
[37]   Wittgenstein, *Remarks on the Philosophy of Psychology*, I, §151. Cf. Wittgenstein, *Zettel*, §545.
[38]   Wittgenstein, 'Philosophy of Psychology – A Fragment,' §1. Cf. Wittgenstein, *Philosophical Investigations*, §650.
[39]   Wittgenstein, *Philosophical Investigations*, §19.
[40]   Ibid., §23.
[41]   Wittgenstein, *Remarks on the Philosophy of Psychology*, II, §15. In all his examples of 'hope' Wittgenstein has 'hope that . . .' in mind. We surely do say of our dogs and cats that they 'hope for' more of whatever treat has been offered.
[42]   Wittgenstein, *Last Writings*, I, §942.

Like non-verbal behaviours or actions, our verbal behaviour is given meaning by the very complex context in which it occurs. Wittgenstein calls this context or background 'the bustle of life'.[43] If we were to produce a study of human behaviour as we do of chimpanzees or giraffes or pelicans, it would have to involve this background which would consist of many things including language, and all that language makes possible:

> How could human behaviour be described? Surely only by showing the actions of a variety of humans, as they are all mixed up together. Not what *one* man is doing *now*, but the whole hurly-burly is the background against which we see an action, and it determines our judgment, our concepts, and our reactions.[44]

Wittgenstein often uses the term 'complicated' to describe aspects of that 'hurly-burly', which includes concepts, play of expressions, behaviour, pattern of life.[45] A metaphor used for this complexity is a tapestry or weaving: '. . . the occasions of grief are interwoven with 1000 other patterns.'[46] We can point to this 'very complicated filigree pattern', recognize it, show it in a novel, but 'can't copy' or completely describe it from the outside, namely, the mind cannot 'encompass' it, to use Diamond's word.[47] It is simpler and much smaller to begin with – that is, in the case of a toddler – but 'bit by bit daily life becomes' more complicated.

> In this case I have used the term, "embedded", have said that hope, belief, etc., were embedded in human life, in all of the situations and reactions which constitute human life.[48]

The complexity of the context in which or the background against which much of our human verbal and non-verbal behaviour occurs is itself created by language. If I see someone putting paper into a slot of some kind, I see a particular action, what '*one* man is doing *now*'.[49] At another level of context I might be 'seeing' someone voting or perhaps mailing a letter. And, 'surrounded by an even more far-reaching manifestation of life',[50] I could be seeing someone voting for a particular candidate or voting without having registered. The occasion and even an entire culture and its political situation might form the context or background needed to determine what is occurring.

---

[43] Wittgenstein, *Remarks on the Philosophy of Psychology*, II, §§625–626.
[44] Ibid., §629.
[45] See, for example, Wittgenstein, *Last Writings*, I, §§876, 946, 967; II, 40e.
[46] Wittgenstein, *Last Writings*, I, §966. Cf. ibid., §862; Wittgenstein, *Zettel*, §568; Wittgenstein, 'Philosophy of Psychology – A Fragment', §§2, 362.
[47] Wittgenstein, *Remarks on the Philosophy of Psychology*, II, §624. For Diamond's use of 'encompass', see her 'The Difficulty of Reality and the Difficulty of Philosophy', 44, quoted earlier.
[48] Wittgenstein, *Remarks on the Philosophy of Psychology*, II, §16; cf. §625.
[49] Wittgenstein, *Zettel*, §567.
[50] Ibid., §534. Cf. Wittgenstein, *Last Writings*, I, §861.

4. Differences of degree

Is there anything astonishing about the possibility of a primitive and a more complicated language-game?[51]

Thanks to extensive research in recent decades we now know just how complex animal behaviour can be as well. This is particularly true of highly social animals. If human behaviour is even more complex due to the fact that so much of it and the background against which it occurs is linguistic, this is a difference of degree and not of kind for Wittgenstein. This can be seen in his use of the term 'primitive' and the fact that there is 'no sharp boundary between primitive forms and more complicated ones'.[52] What are called 'natural expressions' of sensation,[53] of emotion,[54] and even, in some cases, of cognition,[55] behaviours we share with animals and children, are primitive, because they are developmentally there in the beginning:

> But what is the word "primitive" meant to say here? Presumably that this sort of behaviour is *pre-linguistic*: that a language-game is based *on it*, that it is a prototype of a way of thinking and not the result of thought.[56]

> I can easily imagine that a particular primitive behaviour might later develop into a doubt. There is, e.g., a kind of *primitive* investigation. (An ape who tears apart a cigarette, for example. We don't see an intelligent dog do such things.) The mere act of turning an object all around and looking it over is a primitive root of doubt. But there is doubt only when the typical antecedents and consequences of doubt are present.[57]

In the development of the language-game primitive reactions or expressions are replaced by linguistic expressions: 'Primitive pain behaviour is a sensation-behaviour, it gets replaced by a linguistic expression.'[58] As is true in the case of pain, many natural expressions and primitive reactions which form the beginning of a development, may also accompany the later linguistic version and serve as part of the criteria for the application of a concept: 'What is the natural expression of an intention? – Look at a cat when it stalks a bird; or a beast when it wants to escape.'[59]

When it comes to the application of mental terms to animals and humans, levels of development are, again, taken into account both ontogenetically and

[51] Wittgenstein, *Last Writings*, II, 39e.
[52] Wittgenstein, *Philosophical Grammar*, 62.
[53] Wittgenstein, *Philosophical Investigations*, §§256, 244.
[54] Ludwig Wittgenstein, *The Blue and Brown Books* (New York: Harper Torchbooks, 1965), 48.
[55] Wittgenstein, *Remarks on the Philosophy of Psychology*, I, §561.
[56] Wittgenstein, *Zettel*, §541. Cf. Wittgenstein, 'Philosophy of Psychology – A Fragment,' §289.
[57] Wittgenstein, *Remarks on the Philosophy of Psychology*, II, §345.
[58] Wittgenstein, *Remarks on the Philosophy of Psychology*, I, §313. Cf. Wittgenstein, *Philosophical Investigations*, §244; Wittgenstein, 'Philosophy of Psychology – A Fragment,' §289.
[59] Wittgenstein, *Philosophical Investigations*, §647. Cf. Wittgenstein, *Remarks on the Philosophy of Psychology*, I, §862.

phylogenetically: 'The word "read" is applied *differently* when we are speaking of the beginner and of the practised reader'.[60] The contextual and behavioural criteria are different. The concept 'to know' is another example – the criterion for a small child *knowing* how to use a word is not her 'ability to state rules'.[61] About the concept of thinking Wittgenstein says the following:

> We don't say of a table and a chair: "Now they are thinking," nor "Now they are not thinking," nor yet "They never think"; nor do we say it of plants either, nor of fishes; hardly of dogs; only of human beings. And not even of all human beings.[62]

In the case of creatures who don't speak but 'whose *rhythm* of work, play of expression etc. was like our own . . . there is no deciding *how* close the correspondence must be to give us the right to use the concept "thinking" in their case too'.[63] If a monkey only *accidently* puts two sticks together to reach a banana, which would not count as thinking, what might count is what the monkey learns from the accident and how he then goes on to perfect his technique.[64] Here we would have behavioural criteria for the application of concepts like 'considering', 'thinking', 'figuring out'. There are many films now of animals – of ravens, for example – figuring out how to get an out of reach bait in the most ingenious ways. We recognize the context and the behaviour of 'figuring out'.

> So one might distinguish between two chimpanzees with respect to the way in which they work, and say of the one that he is thinking and of the other that he is not.[65]
>
> But here of course we wouldn't have the complete employment of "think." The word would have reference to a mode of behaviour. Not until it finds its particular use in the first person does it acquire the meaning of mental activity.[66]

It is only at the highest levels of development that we find 'the complete employment of "think" '. In the case of the primitive roots of doubt seen in the chimp's investigation of a cigarette, what's missing are the 'typical antecedents and consequences of doubt'.[67] In that case we would not call the chimp's behaviour 'primitive doubting behaviour' but rather one of its roots. What is to be noticed here is that the criteria for the application of mental terms are behavioural and contextual. These terms at their various levels do not refer to mental states or underlying neurology. There is no 'theory' here about what gives us the right to apply a concept. We know the language.

---

[60] Wittgenstein, *Philosophical Investigations*, §156.
[61] Wittgenstein, *Philosophical Grammar*, 62.
[62] Wittgenstein, *Zettel*, §129. Cf. Wittgenstein, *Remarks on the Philosophy of Psychology*, II, §194.
[63] Wittgenstein, *Zettel*, §102. Cf. Wittgenstein, *Remarks on the Philosophy of Psychology*, II, §205.
[64] See Wittgenstein, *Remarks on the Philosophy of Psychology*, II, §224.
[65] Ibid., §229.
[66] Ibid., §230.
[67] Ibid., §345.

Wittgenstein's ambivalence about whether to use the same word for the behaviours in a developmental sequence can be seen in his very extensive treatment of the concept of 'pretence':

> Why can't a dog simulate pain? Is he too honest? Could one teach a dog to simulate pain? Perhaps it is possible to teach him to howl on particular occasions as if he were in pain, even when he is not. But the surroundings which are necessary for this behaviour to be real simulation are missing.[68]

What are those surroundings? A child 'has to learn a complicated pattern of behaviour before he can pretend or be sincere'.[69] Pretending 'is not an experience'.[70] On the preceding page he says the following:

> Can an idiot be too primitive to pretend? He could pretend the way an animal does. And this shows that from here on there are levels of pretence. / There are very simple forms of pretence. / Therefore it is possibly untrue to say that a child has to learn a lot before it can pretend. To do this it must grow, develop, to be sure.[71]

The emphasis on the surroundings – namely, on the complex behaviours, contexts, etc., which make up the criteria for the application of the concept – is meant to counter the mentalist mistake of thinking that pretence is an inner experience of some kind. Once the anti-mentalist point is made, the emphasis is on the development of those 'outer' phenomena, namely, the different degrees of complexity in the background or surroundings, and hence, the different criteria actually used for the application of the concept at various levels. We now have knowledge of animals 'pretending' – squirrels pretending to bury nuts in a certain place when they see that they are being observed by other squirrels, birds pretending to have a broken wing to distract predators from their nest. Of the latter feigning behaviour Jonathan Balcombe writes: '. . . once thought to be merely instinct, recent studies show it to be flexible and calculating.'[72] To say that this isn't 'real' pretending would be like saying that the first grader's reading isn't 'real' reading.

> Could we imagine that people might have a concept of pretence that doesn't coincide with ours? – But would it then be the concept of pretence? – Well, it could be a concept related to ours.[73]

---

[68] Wittgenstein, *Philosophical Investigations*, 3rd edn (Oxford: Blackwell, 1967), §250. Cf. Wittgenstein, *Zettel*, §389; Wittgenstein, *Last Writings*, I, §862.
[69] Wittgenstein, *Last Writings*, I, §869.
[70] Wittgenstein, *Last Writings*, II, 42e.
[71] Ibid., 41e.
[72] Jonathan Balcombe, *Second Nature: The Inner Lives of Animals* (New York: Palgrave Macmillan, 2010), 71.
[73] Wittgenstein, *Last Writings*, I, §224.

But aren't some of the traits of (such) a concept more essential, others less so? That is: If one changes *this* trait it will still be called "pretence" – but if *this* one is changed that word will no longer be used.[74]

5. Other minds, animal and human

We would like to project everything into his inner. We would like to say that *that's* what it's all about. / For in this way we evade the difficulty of describing the *field* of the sentence.[75]

The 'inner' is a delusion. That is: the whole complex of ideas alluded to by this word is like a painted curtain drawn in front of a scene of the actual word use.[76]

A Cartesian mentalist mistake sometimes made by those concerned to combat anthropomorphism is that no matter how great the resemblance between animal and human behaviour is on the 'outside', what goes on on the 'inside' is what our human mental terms really mean. It is thus this inner thing – states of consciousness, experiences, mental processes, etc. – that distinguishes animals and humans and makes the application of certain concepts to animals 'anthropomorphic' and illegitimate. The problem with this assumption is that it leaves us with a scepticism about human minds other than our own. Since, so the assumption goes, all we know by direct observation is our own experience we have to *infer* from the behaviour of the other by analogy with our own case what is going on 'in' him or her. In his *Animal Liberation* Peter Singer makes use of this assumption by pointing out that our lack of certainty about the 'other' in the case of pain, is the same whether the other is an animal or a human and since the behaviour, neurology, etc. is similar, the inference that animals feel pain is perfectly reasonable.[77] To this kind of thinking Wittgenstein has the following retort:

"We *see* emotion." – As opposed to what? – We do not see facial contortions and *make the inference* that he is feeling joy, grief, boredom. We describe the face immediately as sad, radiant, bored, even when we are unable to give any other description of the features. – Grief, one would like to say, is personified in the face. This is essential to what we call "emotion".[78]

I don't *interpret* a facial expression as threatening when someone whips out a knife at me.[79] Nor do I imagine that the feelings of someone 'writhing in pain with evident cause' are hidden from me.[80] There is no inference here.

[74]   Ibid., §225.
[75]   Wittgenstein, *Last Writings*, II, 82e.
[76]   Ibid., 84e.
[77]   Peter Singer, *Animal Liberation* (New York: Harper Perennial, 2009), pp. 10–11.
[78]   Wittgenstein, *Remarks on the Philosophy of Psychology*, II, §570.
[79]   Cf. Wittgenstein, *Zettel*, §218.
[80]   Wittgenstein, 'Philosophy of Psychology – A Fragment,' §324.

This is not the place to rehearse Wittgenstein's complex and thorough dismantling of these assumptions, but rather to see how he brings together adult human and more primitive behaviour to remind us how our language actually works. He imagines an intelligent animal that has been taught to fetch something and take it to another place. The animal starts walking towards the second place without the object and suddenly turns around to fetch the object '*as if it had said* "Oh, I forgot . . .!" ' [81]

> If we were to see something like this we would say that at that time something had happened within it, in its mind. What then has happened within *me* when I act this way? "Not much at all," I would like to say. And what happens inside is no more important than what can happen outside, through speaking, drawing etc. (From which you can learn how the word "thinking" is used.) [82]

Does a cat waiting by a mouse hole and an experienced robber waiting for his victim have to be thinking about the victim in question for us to see that they are lying in wait for them? [83] No, we know the contexts and behaviours of 'lying in wait for . . .' in both cases.

The radical Cartesian scepticism about other minds that Wittgenstein is concerned to combat takes the form of 'we can *never* know . . .' In the case of 'pretence' his response to this is the following:

> But what does it mean to say that all behaviour *might* always be pretence? Has experience taught us this? How else can we be instructed about pretence? No, it is a remark about the concept 'pretence'. But then this concept would be unusable, for pretending would have no criteria in behaviour. [84]

This kind of scepticism results from a temptation 'to misunderstand the logic of our expressions' [85] and is distinguished from what Wittgenstein calls 'ordinary uncertainty' and 'philosophical uncertainty'. Ordinary uncertainty relates to particular cases and is 'practical and primitive'. [86] Moreover, it 'is an (essential) trait of all these language-games'. [87] In certain circumstances it makes sense to wonder whether someone is pretending and in others it doesn't. Philosophical uncertainty, on the other hand, 'relates . . . to the method, to the rules of evidence', [88] namely, to the question whether the criteria for the application of a concept actually exist. The belief that a particular person is not in pain because he might be pretending has 'different *grounds*' or 'a

[81] Wittgenstein, *Remarks on the Philosophy of Psychology*, II, §6.
[82] Ibid. Cf. Wittgenstein, *Last Writings*, I, §828.
[83] Wittgenstein, *Remarks on the Philosophy of Psychology*, I, §829.
[84] Wittgenstein, *Zettel*, §571. Cf. Wittgenstein, *Philosophical Investigations*, §345.
[85] Wittgenstein, *Philosophical Investigations*, §345.
[86] Wittgenstein, *Remarks on the Philosophy of Psychology*, II, §558.
[87] Wittgenstein, *Last Writings*, I, §877.
[88] Wittgenstein, *Remarks on the Philosophy of Psychology*, II, §682.

different logic' from the belief that an amoeba feels no pain.[89] This is brought out further in the following:

> Think of the uncertainty about whether animals, particularly lower animals, such as flies, feel pain. / The uncertainty whether a fly feels pain is philosophical; but couldn't it also be instinctive? And how would that come about? / Indeed, aren't we really uncertain in our behaviour towards animals? One doesn't know: Is he being cruel or not.[90]
>
> For there *is* uncertainty of behaviour which doesn't stem from uncertainty in thought.[91]

As we've seen above, the concept 'pain' 'seems able to get a foothold' in the case of a wriggling fly but not in the case of a stone. Even so, we can, in our thought, be uncertain whether the wriggling sufficiently resembles the pain behaviour we know in humans and 'higher' animals and thus provides criteria for the application of the concept: 'If however I doubt whether a spider feels pain, it is not because I don't know what to expect.'[92] This is philosophical uncertainty. On the other hand, I could also have an instinctive reaction to this particular fly's wriggling when I notice its torn wing.

When it comes to the pain of other human beings, not only are we not Cartesian sceptics, but we are also hard-wired to respond: 'Being sure that someone is in pain, doubting whether he is, and so on, are so many natural, instinctive, kinds of behaviour towards other human beings . . .'[93] In fact, it is a 'primitive reaction' to respond to someone else's pain.[94]

What about the case of those animals that resemble us the least? Wittgenstein imagines how we might treat someone if we didn't believe the other feels pain – 'as lifeless, or as many treat those animals that least resemble humans. (Jellyfish, for instance.)'[95] When animals don't resemble us enough, our natural empathy is not activated – unless we have the help of the painstaking research that allows us to consider the possibility that fish, for example, feel pain.[96] In some cases it might even produce our natural reaction to the pain of others. As a child, when I was being taught how to catch and clean fish, I felt some amusement but basically great squeamishness at the flipflopping of a just caught fish. Now that I know fish are suffocating to death when they behave like that out of water, I am able to see the struggle to breathe and the look in the eyes. I am able to see the resemblance to us in similar circumstances, in a lethal habitat; in our case, being forcibly held under water. So much research done since Wittgenstein's time helps us recognize in the behaviour of animals all kinds of things

---

[89]   Wittgenstein, *Last Writings*, I, §242; ibid., §239, fn. 1.
[90]   Wittgenstein, *Remarks on the Philosophy of Psychology*, II, §659.
[91]   Ibid., §660.
[92]   Wittgenstein, *Zettel*, §564.
[93]   Ibid., §545.
[94]   Ibid., §540.
[95]   Wittgenstein, *Last Writings*, §238.
[96]   See Victoria Braithwaite, *Do Fish Feel Pain?* (Oxford: Oxford University Press, 2010).

we otherwise might not see, and thus enables us to remove the uncertainty he calls 'philosophical'. Although Wittgenstein assumed that fish, for example, do not 'think',[97] had he known what we now know about the intelligence of fish,[98] he might be the first to change his mind. It isn't that we are discovering new criteria for our concepts, but rather discovering contexts and behaviours we didn't see before and which we recognize as the criteria for concepts like 'intelligent' or 'think'. Again, whether we apply the concept or not depends on which of the concept's essential traits are present in the observed behaviour. In the absence of certain of the usual criteria, the question may arise whether it is then a primitive version of the concept or a different concept altogether?[99]

When it comes to suffering, what about the animals that do resemble us? Here, unlike the case of 'lower' animals where we are not sure, our natural reaction is to see the suffering of the animal immediately, not by inference. Wittgenstein seems quite certain about the suffering in a bullfight:

> In a bullfight the bull is the hero of a tragedy. Driven mad first by suffering, he then dies a slow and terrible death.[100]

6. Nothing lacking

> The difficulty of renouncing all theory: One has to regard what appears so obviously incomplete, as something complete.[101]

Though there are differences between animals and humans due, among other things, to the nature of human language, animals are not seen by Wittgenstein to be *lacking* anything human beings have. There are two ways to see this. One is in the distinction between two kinds of statement. To say that an eagle *can't* fly because it has a broken wing is to make an empirical statement about a failure or a lack of a capacity of an eagle. Even though they have wings, penguins can't fly, and although we would not call this a failure, we *might* call it a lack, since they are birds and almost all birds have the capacity to fly. To say, on the other hand, that worms or humans *can't* fly is to make what Wittgenstein calls a 'grammatical' statement, namely, a statement about the 'logic of [our] language'.[102] In this case it says that both the concept 'to fly' and its opposite are not applicable to these animals. Worms and humans are not lacking something, because they never had the ability to fly in the first place. To see this as a lack is to

---

[97] Wittgenstein, *Zettel*, §117.
[98] See Balcombe, *Second Nature*, 41–2, 86, 100, 110, 115 and elsewhere.
[99] Cf. Wittgenstein, *Last Writings*, I, §225.
[100] Ludwig Wittgenstein, *Culture and Value* (Chicago, IL: University of Chicago Press, 1980), 50e.
[101] Wittgenstein, *Remarks on the Philosophy of Psychology*, I, §723.
[102] See, for example, Wittgenstein, *Philosophical Investigations*, §93. Wittgenstein often uses the terms 'grammar' and 'logic' interchangeably in his later work; see, for example, *On Certainty*, §§56–7, et passim.

project onto them the denial of a concept that doesn't apply at all in either its positive or negative form. On the other hand, to say that worms and horses 'can't' talk the way we humans do and to mistake this for an empirical 'can't' is to see these animals as lacking something, when in fact they are lacking nothing at all. They are what they are. Ironically, to apply the denial of a concept that doesn't apply in the first place turns out to be a form of 'anthropomorphism'.

What Wittgenstein is pointing to here are the criteria for the application of certain concepts, not certain empirical issues. This has everything to do with the word, 'can't'.

> Why can a dog feel fear but not remorse? Would it be right to say "Because he can't talk"?[103]

> Only someone who can reflect on the past can repent. But that does not mean that as a matter of empirical fact only such a one is capable of the feeling of remorse.[104]

> But a machine surely cannot think! – Is that an empirical statement? No. We say only of a human being and what is like one that it thinks.[105]

The grammatical 'cannot' means that the rules of the game forbid this kind of move. In certain cases like that of describing some differences between human beings and the various animals, it is easy to mistake the grammatical 'cannot' for an empirical one.[106] Wittgenstein, therefore, suggests, 'Do not say "one cannot", but say instead: "it doesn't exist in this game".'[107] 'A dog cannot be a hypocrite; but neither is it sincere.'[108] Neither of these opposite concepts applies to dogs. When Wittgenstein says, as he often does, 'We say . . .' or 'We do not say . . .' this, again, refers to the 'rules of the game', to what we 'can' and 'cannot' say. It is not only a question of what various animals and humans, and even tables and machines, can and cannot do, but also of what concepts we speakers can and cannot apply. We cannot (meaningfully) say that the number 7 is colourless – it is neither coloured nor colourless. Nor can we meaningfully say, 'She's in pain, but she's showing it.' The rules of the game forbid it. In the empirical sense we can (are able to) say what we please, but that doesn't automatically give what we say sense. What Wittgenstein sees in our language allows for new concepts and old ones that change, but 'it presupposes that most concepts remain unaltered'.[109] It is the most basic concepts he investigates in his later work. 'Could a legislator abolish the concept of pain? / The basic concepts are interwoven so closely with what is most fundamental in our way of living that they are therefore unassailable.'[110]

---

[103]   Wittgenstein, *Zettel*, §518.
[104]   Ibid., §519.
[105]   Wittgenstein, *Philosophical Investigations*, §360; cf. §344.
[106]   This is also Wittgenstein's criticism of metaphysics. See, for example, Wittgenstein, *Zettel*, §458.
[107]   Wittgenstein, *Zettel*, §134.
[108]   Wittgenstein, *Last Writings*, I, §870.
[109]   Wittgenstein, *Last Writings*, II, 43e.
[110]   Ibid., 43e–44e.

The other way in which the fact that animals lack nothing we humans have is brought out with the notions of 'natural history' and 'completeness':

It is sometimes said: animals do not talk because they lack the mental abilities. And this means: "They do not think, and that is why they do not talk." But – they simply do not talk. Or better: they do not use language – if we disregard the most primitive forms of language. – Giving orders, asking questions, telling stories, having a chat, are as much a part of our natural history as walking, eating, drinking, playing.[111]

Human natural history includes, among other things, various linguistic activities; the natural history of bats includes echolocation; and that of mockingbirds, flying and perfect imitation of other birdsongs. Animals simply do not talk and we humans simply do not live under water, hear sounds above a certain pitch, or fly.

'If I tell someone "Men think, feel, . . .", it seems I am making a statement of *natural history* to him. It might be intended to show him something about the difference between man and the various kinds of animals.'[112] Of course, Wittgenstein has shown us how more 'primitive' versions of concepts like 'think' do apply to certain animals. In the case of measuring,[113] and going through a proof and accepting its results,[114] both mentioned as facts of our human natural history, only the latter seems to be strictly human. There may be some animals that exhibit something like measuring behaviour, I don't know.

When it comes to language, primitive forms are seen in both animals and children. In children we see the primitive form of *our* language. In young bees, dolphins, or elephants, no doubt we would see the primitive form of *their* respective forms of expression and communication. (I have seen and heard a very young mockingbird just beginning to mimic the songs of others.) Not only is there 'no sharp boundary between primitive forms and more complicated ones',[115] but more primitive forms are complete in themselves, whether we are thinking ontogenetically or phylogenetically:

The primitive language-game which children are taught needs no justification; attempts at justification must be rejected.[116]

I want to regard man here as an animal; as a primitive being to which one grants instinct but not ratiocination. As a creature in a primitive state. Any logic good enough for a primitive means of communication needs no apology from us. Language did not emerge from some kind of ratiocination.[117]

[111] Wittgenstein, *Philosophical Investigations*, §25.
[112] Wittgenstein, *Remarks on the Philosophy of Psychology*, II, §18.
[113] Wittgenstein, *Remarks on the Philosophy of Psychology*, I, §1109.
[114] Ludwig Wittgenstein, *Remarks on the Foundations of Mathematics*, 3rd edn (Oxford: Blackwell, 1978), pt. 1, §63.
[115] Wittgenstein, *Philosophical Grammar*, 62.
[116] Wittgenstein, 'Philosophy of Psychology – A Fragment,' §161.
[117] Wittgenstein, *On Certainty*, §475.

A very simple language imagined at the beginning of *Philosophical Investigations* Wittgenstein asks us to conceive of 'as a complete primitive language'.[118] It consists of three words – 'slab', 'block', 'beam' – called out by builders. One way the completeness is shown is by asking whether the word 'slab' in the builders' language has the same meaning as the word 'slab' in our language. The answer is no, because in addition to context and matching up the right word with the right object, a large part of what determines a word's meaning or makes a concept what it is are its connections to other words, perhaps even the rest of the language. Wittgenstein has a metaphor for this: ' "I set the brake up by connecting rod and lever." – Yes, given the whole of the rest of the mechanism. Only in conjunction with that is it a brake-lever, and separated from its support it is not even a lever; it may be anything, or nothing.'[119] Because the 'rest of the mechanism' is so different in these two languages, due to size and complexity, the word 'slab' could not have the same meaning. If we assumed that it did by assuming that words are not conceptually connected, then the primitive language would not be whole or complete, it would be lacking many 'parts'. The following is another example of this point:

> A child who learns the first primitive verbal expression for its own pain – and then begins (also) to talk about his past pains – can say one fine day: "When I get a pain the doctor comes". Now has the word "pain" changed its meaning during this learning process? – Yes, its use has changed. / But doesn't the word in the primitive expression and the word in the sentence refer to the *same thing*, namely, the same feeling? To be sure; but not to the same techniques.[120]

It is not just animals and children who should not be considered as lacking something normal adult human beings have:

> What would a society all of deaf men be like? Or a society of the 'feeble-minded'? *An important question!* What then of a society that never played many of our customary language-games?[121]

> One imagines the feeble-minded under the aspect of the degenerate, the essentially incomplete, as it were in tatters. And so under that of disorder instead of a more primitive order (which would be a far more fruitful way of looking at them). / We just don't see a *society* of such people.[122]

Not seeing a 'society' of such people prevents us from seeing the completeness, the wholeness of the deaf or the feebleminded and, instead, makes us think of

---

[118]   Wittgenstein, *Philosophical Investigations*, §2; cf. §18.
[119]   Ibid., §6.
[120]   Wittgenstein, *Last Writings*, I, §899. Cf. Wittgenstein, *Philosophical Investigations*, §555.
[121]   Wittgenstein, *Zettel*, §371.
[122]   Ibid., §372.

them as lacking something, just as some see animals as lacking something. But, as Wittgenstein reminds us,

> A treatise on pomology may be called incomplete if there exist kinds of apples which it doesn't mention. Here we have a standard of completeness in nature. Suppose on the other hand there was a game resembling that of chess but simpler, no pawns being used in it. Should we call this game incomplete?[123]
>
> ... a colour-blind man is in the same situation as we are, his colours form just as complete a system as ours do; he doesn't see any gaps where the remaining colours belong.[124]

Interestingly, Wittgenstein says that he would not say of himself or anyone else that 'we understood manifestations of life that are foreign to us. And here, of course, there are degrees'.[125] He makes it clear that this is not an empirical statement. Right above this paragraph he distinguishes between 'observing' and 'participating' in life and seems to use the life of fish and of plants as examples of what can only be observed: 'We can't talk about the joy and sorrow, etc., of fish.'[126] With domestic animals, especially pets we live with, there is a great deal of overlap of our lives and true participation. I think the reason Wittgenstein famously said that 'If a lion could talk, we could not understand him,'[127] is that he couldn't imagine that overlap in the case of a lion. Anyone who has seen the various documentaries of how adult lions and other great cats relate to those humans who raised them as cubs can easily see that overlap. What is striking is Wittgenstein's respect for the wholeness of the lives of other species whether we understand them or not.

Another important aspect of natural history and completeness are the primitive roots and natural expressions that form the ground or foundation necessary for what can be learned or taught or trained. As we've seen, Wittgenstein reminds us that pointing will form an important part of early language learning 'because it is so with human beings; not because it could not be imagined otherwise'.[128] Another example he gives us is that part of the natural history of dogs but not cats is that they can be taught to retrieve.[129] At the level at which training as opposed to explaining is taking place there has to be some capacity for what is being taught, and a natural tendency to respond to what is being shown by the teacher and to the kinds of encouragement being given. A mouse can't be taught to herd sheep or a chipmunk to heel or speak. Nor can a child be taught how to fly. None of these creatures taken as the subjects of grammatical 'can't'-statements is lacking something other creatures have.

---

[123]   Wittgenstein, *The Blue and Brown Books*, 19.
[124]   'Wittgenstein, *Zettel*, §257.
[125]   Wittgenstein, *Remarks on the Philosophy of Psychology*, II, §30.
[126]   Ibid., §29.
[127]   Wittgenstein, *Philosophical Investigations*, 3rd edn, pt. 2, §xi, 223e.
[128]   Wittgenstein, *Philosophical Investigations*, 4th edn, §6.
[129]   Wittgenstein, *The Blue and Brown Books*, 89–90; Wittgenstein, *Zettel*, §§186–7.

7. Whose 'anthropomorphism'?

The charge of 'anthropomorphism' is the charge that terms appropriate only to human beings are being projected onto animals. Other than cases of gross sentimentality or children's books, it is hard to see why much of our ordinary talk about animals is 'anthropomorphic'. *Waiting, being patient, thinking, being in pain, caring for young, playing, wanting more, being afraid, asking to go out, recognizing someone, knowing how, being happy, clever, sad, solving a problem, engaging in deception, recognizing oneself in a mirror, etc., etc.* are all terms commonly applied to animals in nature, on the farm, in the home, in veterinary medicine and in the laboratory. We language users know the behaviour and the contexts which provide the criteria for the application of such concepts. We also know that the criteria for the same concept differ at different levels of development. We know, for example, the criteria for the application of the concept 'recognize' in the case of an amoeba's *recognition* of a nutrient as nutrient, and the more complex criteria for the case of a dog's *recognition* of his master, or a chimp's *recognition* of a new sign; and we know that there are much more complex criteria for the case of a botanist's *recognition* of a new species of orchid. There is, of course, a different neurology underlying each of these cases, and different cognitive steps leading up to the occurrence of recognition, but these have nothing to do with the criteria for the application of the concept of 'recognition'.

The ancient Greeks knew nothing of neurology or cognitive psychology and had no more difficulty applying the concept of 'recognition' than we do. After many years Homer's Odysseus returns home. He is disguised and his dog, Argos, is in very bad shape, but they both recognize one another. For the scientist to be interested in the underlying neurology or the cognitive processes accompanying recognition, she already has to know what recognition is and how to apply the concept. We humans do not have the same neurology as bats, octopuses or chimpanzees, but the underlying neurology or cognition is not our problem here. It is rather the meaning of our words, particularly our mental words whose use requires, as Diamond puts it, a 'sense of our own bodily life and our capacity to respond to and imagine the bodily life of others'.[130] It is the criteria for the application of certain concepts.

To see animals as lacking something human beings have is the true projection. The world of any individual species is complete in itself. That we humans 'cannot' even approach many of the capabilities of various animals we have never considered to be a lack. Why consider their world to be lacking something ours has? This is true anthropomorphism. If the projection onto animals of terms taken to be applicable only to humans is anthropomorphism, equally 'anthropomorphic' is the projection onto animals of the denial of those terms.

---

[130]   Diamond, 'The Difficulty of Reality and the Difficulty of Philosophy', 53.

4

Rape among Scorpionflies, Spouse Abuse among the Praying Mantis and Other 'Reproductive Strategies' in the Animal and Human World

Olli Lagerspetz

Whatever else we may be, we are also animals. On the whole it should be a good thing for theories of human behaviour to look for continuities between human life and other life forms. But there will be questions about how far and where exactly we should go. Exaggerated emphasis on human uniqueness may squander important opportunities for reflecting on our animal nature. But if we do seize those opportunities we run the twin risks of anthropomorphism and biologism. We may illegitimately attribute human traits to animals or we may, on the contrary, too easily assume that specifically human traits are reducible to biology. But there is also a third risk: that of misbegotten rationalism, blinding us to important features of life, both human and animal.

One familiar form of rationalism involves the tendency always to look for conscious reasoning behind people's behaviour and choices. Moreover, rationalism often represents reasoning as a rather narrow kind of calculus from ends to means and vice versa. But rationalism may also mean that various basic traits of human or animal life are treated *as if* they resulted from reasoning. An attitude of this kind means that we expect explanations of behaviour always to make reference to something that has the form of rational thought processes, even in cases which are admittedly not instances of rational purposive action. This kind of approach is characteristic of many of the explanations currently put forward within evolutionary psychology. The occurrence of particular behavioural traits in humans is explained by showing in what ways they can be seen as rational means to achieving successful reproduction in our evolutionary past, although it is not implied that the individual, or anyone else, acts or has ever acted *in order to* reproduce. While such explanatory attempts should not be ruled out immediately, they do invite questions about their own role as explanations. In what sense can they claim to unearth real motivating forces behind behaviour? As we will see, related questions have already for a long time been discussed in the philosophy of the social sciences, outside the context of the present debate.

1. Evolutionary psychology: Proximate versus ultimate levels of explanation

Evolutionary psychology is defined as the study of how human thinking and behaviour have been shaped by evolution.[1] But it is also committed to a specific theoretical approach. Evolutionary psychology aims to explain existing patterns of human and animal behaviour in terms of their contribution to increased fitness over evolutionary time, relative to any other behaviour in the same behavioural domain.[2] To put it simply, certain behavioural dispositions of individuals or populations are results of natural and sexual selection which have favoured traits that, in the past, contributed to the successful reproduction of their ancestors.

Just as economic theory does not necessarily presuppose greed as the dominant human character trait, neither does evolutionary psychology assume any conscious wish to reproduce. Theoretically careful evolutionary psychologists state this explicitly.[3] A reader of popular work and basic introductory textbooks on evolutionary psychology may find that statement surprising, given the proliferation of talk suggesting that animals are more or less obsessed with eugenics. A textbook by Matthew Rossano is a case in point. This is Rossano commenting on insect mating:

> [T]he female hanging fly refuses sex unless she is fed a dead insect during the act. Even then, if the meal is so skimpy that she can finish it before he finishes with her, she simply moves on to another. She *apparently does not want* the genes of someone who does not have the wherewithal to scrounge up at least one decent dinner.[4]

The quote seemingly tells us that female insects consciously look for partners with the right genes.[5] The same work also presents things as if *genes* could carry out reasoning or make judgements.[6] But anthropomorphic depictions are used

---

[1] Matthew J. Rossano, *Evolutionary Psychology: The Science of Human Behavior and Evolution* (Hoboken, NJ: Wiley, 2003), 23–4.

[2] See ibid., 38–9, 55.

[3] '[R]esearchers using an evolutionary psychological perspective often frame hypotheses in terms of the costs and benefits to an organism of performing a particular behaviour. These costs and benefits refer to the effects on reproductive success over evolutionary time. Costs decrease the probability of successful reproduction, whereas benefits increase the probability of successful reproduction. These terms are sometimes misconstrued as referring to a more general idea of perceived costs and benefits to the individual or to society. However, these terms carry no moral or ethical meaning and are used only in terms of naturally selected biological functioning' (William F. McKibbin et al., 'Why Do Men Rape? An Evolutionary Psychological Perspective,' *Review of General Psychology*, 12 (2008), 87).

[4] Rossano, *Evolutionary Psychology*, 224 (emphasis added). Rossano is citing data from Randy Thornhill's 'Sexual Selection and Paternal Investment in Insects,' *American Naturalist*, 110 (1976), 153–63.

[5] See also Rossano, *Evolutionary Psychology*, 224, where we learnt that the female will not mate 'with a weak, diseased, or dim-witted male' as it 'might mean wasting her precious time, eggs, and energy on offspring unlikely to survive to viability'.

[6] See ibid., 273: 'hominin females . . . looked for outward displays of quality, some of which are as obvious *to a woman's genes* as the nose on man's face'; and 283: 'the best way that *genes can deal with* the world's imperfections is to be ready to *use* a mix of reproductive strategies' (all emphases are mine); cf. 244.

there for illustrative purposes only, not as literal attempts to describe animal reasoning.

Evolutionary explanations in terms of fitness enhancement certainly do not claim to spell out the motives or reasoning that guide the behaviour of individuals. The claim is rather that motives and reasons in the everyday sense should not be seen as the final explanation. Behind such *proximate* causes of behaviour, another explanation can be provided at a deeper, or *ultimate*, level – and the concerns that guide us at *that* level should be described in terms of fitness enhancement.[7]

Nevertheless, explanations in terms of fitness enhancement have a form that somehow connects behaviour with an assumed intelligible purpose. While I am not saying that such 'ultimate' explanations are no explanations at all, there will be questions about precisely how they connect with explanations in terms of motives, purposes and judgements in any ordinary sense. Most worryingly, by making use of such language, the 'ultimate' explanations give the impression of substituting and falsifying other, 'proximate' explanations that make genuine references to motives, purposes and judgements. The question whether the proposed ultimate explanations are factually right (i.e. whether they describe evolutionary mechanisms that have truly influenced behavioural dispositions in the proposed ways) falls outside the scope of the present essay as well as my qualifications. Instead, I will focus on the question: What is the relation between the proposed explanations and claims about motives for action?

By way of example, two cases of insect behaviour will be discussed. In the literature, they are presented as instances of reproductive strategies. My point is, first, that the description of these behavioural patterns as *strategies* is based on a deductive argument that, despite appearances, requires no reference at all to animal psychology. Secondly, this means that they should only be called 'strategies' once it is understood that the word, in this context, is quite devoid of its usual association with purposive rational action. Thirdly, it follows that arguments of this kind cannot claim to have unearthed the 'real' purposes or hidden rationality behind human or animal action. What evolutionary psychologists *can* claim is to identify evolutionary conditions, such as selective pressures, under which certain kinds of predisposition have become *viable* and hence survived (assuming that they have a hereditary component).

This is not to say that there is no point in drawing analogies between human and animal life. One alternative approach, which will be no more than hinted at here, is to recognize continuities between human beings and other animals, but to do so at the level of our common lived experience.

2. The praying mantis: Complicity or conflict?

The praying mantis is a cricket-like insect known for its (humanly speaking) bizarre mating habits. Copulation frequently ends with the female devouring the male. There

---

[7]   Ibid., 25–6.

is some indication that this behaviour is less frequent in nature than in captivity, but at least *Mantis religiosa*, which is one among the several existing species of praying mantis, apparently cannot mate properly even in natural conditions unless the male's head is bitten off in the process. The beheaded male's copulatory movements continue and in fact intensify, leading to fertilization. Thanks to the extra nourishment, the female will be able to produce a better protective case in which the eggs mature, and she will have more energy to guard the case against predators.

Researchers disagree about whether being eaten up is in the 'interest' of the *male* mantis. The question may seem odd. Ending up as food on someone else's plate may look like an unqualifiedly grim prospect. However, from a different angle, the answer is not fully clear. Decapitation by one's partner excludes the prospect of further matings. But if procreation is the goal, sacrificing one's life may yet be the 'rational' thing to do:

> One of the original theoretical models of sexual cannibalism shows that a male should be willing to sacrifice his life to an inseminated partner if he can expect little subsequent mating and if his value as a food item would allow the female to rear substantially more offspring.[8]

Does the male 'willingly' sacrifice his life? There is a difference between scientists on how to argue on the issue. The question has been framed as one of 'complicity or conflict'.[9] Some authors would study the males' behaviour in order to see whether their behaviour suggests that they try to avoid being killed during copulation – which appears to be the case.[10] These researchers conclude that there is a *conflict*. Others analyse, instead, the outcome and argue it is to the male's *reproductive advantage* to be killed. They infer there is *complicity*. Thus they do not take the male's behaviour to settle the issue.

The disagreement is about *how to identify* motivating forces and interests in the context of animal behaviour. What should be the starting point: what the animals apparently think, or how their behaviour fits into a larger scheme of reproductive effort? The resolution of the scientific disagreement is, in other words, not entirely dependent on empirical data.

3. The concept of reproductive strategy

The initial example may serve to introduce four basic concepts: *fitness, mating effort, parental investment* and *reproductive strategy*. These concepts originate in ecology and evolutionary biology, but now they have become theoretical cornerstones of evolutionary psychology.

---

[8] Jonathan P. Lelito and William D. Brown, 'Complicity or Conflict over Sexual Cannibalism? Male Risk Taking in the Praying Mantis *Tenodera aridifolia sinensis*,' *American Naturalist*, 168 (2006), 263–9, at 263.
[9] Ibid., 264.
[10] Ibid., 268.

By an individual's *fitness*, evolutionary biologists mean the number of alleles (variants of genes that influence the individual's phenotype, creating individual differences between animals) transmitted further over generations. This may happen either through the individual's own progeny or via the progeny of other biological kin with whom the individual shares alleles by common descent.[11] Thus the *individual's* welfare influences its fitness only to the extent that it increases the individual's capacity to advance the spread and survival of its alleles. As the quote above suggests, the male mantis may increase his 'fitness' by mating and then being consumed by his partner.

The individual may advance the spread and survival of its alleles in two ways. It may do so directly – by actively seeking mating opportunities (mating effort) – or indirectly, that is, by behaving in ways that enhance the fitness of its offspring and other kin.

Mating effort is defined as the amount of resources, such as time and energy, that the individual spends on acquiring mates.[12] *Parental investment* (or parental effort) is the amount of resources spent by the individual on helping its offspring to survive. It includes resources spent, for instance, on feeding and bringing up children, but also on the production of viable sperm and egg cells.[13] The behaviour of the male mantis – offering up himself as food – can be said to represent parental investment in his future offspring.

Various species, but even individuals within the same population, may have divergent *reproductive strategies*; that is, different behavioural patterns with regard to their utilization of resources in reproduction. A maximally successful reproductive strategy is, by definition, one that maximizes the individual's (inclusive) fitness given the amount of available resources.

We should note that the reasoning of evolutionary biologists that I have so far outlined is simply deductive, starting out from certain definitions. The argument can be reconstructed as follows:

1. (*Def.*) The amount of resources available to each individual is limited.
2. (*Def.*) Different patterns of behaviour require different quantities of resources.
3. (*Def.*) Different patterns of behaviour result in different quantities of surviving alleles in the next generation.
4. (*Def.*) Patterns of behaviour can be compared in terms of their cost efficiency. They involve different ratios between input, in the form of resources, and output, in the form of surviving alleles. They can be summed up as *reproductive strategies*.
5. It follows by definition from 1 to 4 that, other things being equal, individuals with more cost-efficient reproductive strategies end up with a higher number of surviving alleles relative to the population; in other words, with higher fitness.
6. Since 'fitness' is measured in the proportionate, not absolute, amount of the individual's alleles in the next generation, each increase in one individual's fitness

---

[11]   William D. Hamilton, 'The Genetical Evolution of Social Behavior: I and II,' *Journal of Theoretical Biology*, 7 (1964), 1–52.
[12]   Randy Thornhill and Craig T. Palmer, *A Natural History of Rape: Biological Bases of Sexual Coercion* (Cambridge, MA: MIT Press, 2000), 33.
[13]   Ibid.

for the most part diminishes that of the others.[14] In this sense, individuals may be described as *competing for fitness*.

7. Insofar as the relevant behavioural dispositions are hereditary, other things being equal, some of those patterns will be propagated over time more than others in the population.

The argument so far involves no assumptions about what animals think or want. The model does not say that animals subjectively see their use of resources as a cost; for instance, that they see childcare as a strain that they wish to avoid. Nor does it say that they experience a wish to maximize their progeny. Different patterns of behaviour simply lead to different outcomes in terms of relative amounts of surviving alleles, and the assumption is made that these patterns of behaviour are based on innate predispositions of the animal.

The assumption of universal competition in this argument does not imply competitiveness as a universal psychological trait, with its implications of keeping track of one's performance and comparing it with that of others. If indeed it did, it would amount to anthropomorphism no less staggering than if we were to attribute vanity to peacocks or musical interests to nightingales. It just means that behaviour results in outcomes that may be *ranked, by scientists*, according to some standard. Quite analogously, spheres rolling down a slope might be ranked according to which of them reaches the bottom first; the spheres might be described as 'racing' or 'competing'.

Being deductive, reasoning of this type can be applied to any living beings, including insects and humans, even to computer programmes. The approach is reminiscent of – and historically derived from – rationalist political philosophy, notably the work of Thomas Hobbes. Hobbes argued that, given the scarcity of resources and given certain assumptions about behaviour, which he deductively derived from the concept of motivation, human interaction in the absence of coercive institutions will necessarily turn into relentless competition for resources.[15] Such approaches were subsequently refined and formalized in economic theory and game theory.[16]

4. Rape as a reproductive strategy

In principle, two reproductive strategies exist in the animal kingdom. Simplistically, the contrast can be described as one of quantity versus quality.[17] One is to beget as many offspring as possible – that is, high mating effort. But it naturally follows that the parental investment expended on each of the numerous young will generally be

---

14   This does not hold when the individuals are close kin or when they reproduce together. In the latter case, the effects of mating on the mate's fitness will depend, among other things, on how it influences its prospects for further matings.

15   Thomas Hobbes, *Leviathan* (London: Penguin, 1985 [1651]).

16   Ann Cudd, 'Game Theory and the History of Ideas about Rationality: An Introductory Survey', *Economics and Philosophy*, 9 (1993), 101–33.

17   Rossano, *Evolutionary Psychology*, 223.

low. The other option involves fewer young, making sure as many of them as possible survive to reproduce – that is, high parental investment.

Building on Robert Trivers' theoretical work, evolutionary psychologists emphasize the role of parental investment for sexual selection.[18] In species that reproduce themselves sexually, each individual offspring has two parents. Thus parental investment by one parent of the shared offspring also contributes to the fitness of the other parent. This means that, in competing for a mate, the animals also compete for access to the parental investment by their prospective mate. As Randy Thornhill and Craig Palmer put it:

> In the terminology of evolutionary biology, parental effort is a *resource* that an individual can *possess* and that other individuals *desire*. . . . Indeed, from an evolutionary perspective, parental effort is the essential resource, because it determines how many offspring there will be and their likelihood of surviving. The usual way of obtaining this resource from another individual is through sexual copulation, in which one individual's parental investment is, in a sense, "taken" by another and used by that individual to produce its offspring.[19]

Once more the argument is deductive. It is reminiscent of economic theory because the situation is described as competition for a desired resource, and the task facing each individual is how to secure a maximal share with a minimal investment.[20] Thus, continuing the deduction above:

8. Increase of parental investment spent on the individual's offspring tends to increase the individual's fitness.
9. Assume that the individual, A, and its mating partner, B, have shared offspring. By definition, parental investment by B on the shared offspring also contributes to A's fitness (and vice versa).
10. B's parental investment on any other offspring (by individuals genetically unrelated to A) will not contribute to A's fitness (nor vice versa).
11. Hence (*from 6, 8–10*), A *competes* with other individuals for *access* to B's parental investment.

This analysis of the argument highlights the fact that 'competition' and 'desire' have no psychological content here. The concluding point merely states that a certain outcome tends to enhance the individual's fitness. Increased parental investment by B will *make it more likely* for A's alleles to survive.

Thus far, the situations of A and B have been described as symmetrical. However, the explanatory framework of evolutionary psychology is strongly influenced by the

[18] Robert L. Trivers, 'Parental Investment and Sexual Selection,' in *Sexual Selection and the Descent of Man: 1871–1971*, ed. Bernard Campbell (Chicago, IL: Aldine, 1972), 136–79. See also Thornhill and Palmer, *A Natural History of Rape*, 33f.
[19] Thornhill and Palmer, *A Natural History of Rape*, 34.
[20] See Rossano's *Evolutionary Psychology*, 222–3, where 'business' and 'market' are used as metaphors.

fact that asymmetry rather than symmetry is the rule. In most species that reproduce themselves sexually, males and females of the same species can be rationally expected to follow divergent reproductive strategies. In brief, one may expect more mating effort by males and more parental investment by females.

The number of offspring by an individual male tends to increase in proportion with the number of matings, while this is true of females only up to a limit. Given the fact of female pregnancy, for viviparous animals (animals that give birth rather than laying eggs) there is a ceiling to how many young an individual female may beget. This is also true generally of species whose egg cells are considerably larger and fewer in number than their male sperm. Each pregnancy and each fertilized egg already represents considerable parental investment by the female, but not by the male.

On the other hand, there is no fixed limit to how many young may be sired by a male. Hence, it generally lies in the reproductive interest of the male to inseminate as many females as possible but *not* to engage in additional parental investment unless he can reconcile it with keeping up his mating activities elsewhere. The implication is that mating will often lie in the reproductive interest of the male but not of the female.[21]

This places females in a position where they can be selective of sexual partners. They can choose among males who compete with one another for access to their parental investment.[22] The females' fitness is enhanced if they mate with males who are either genetically well-endowed (measured in terms of their general chances of begetting vigorous young that survive to reproductive age) or who contribute to the care and protection of the shared offspring.[23]

The idea of divergent reproductive strategies provided the theoretical backbone for Thornhill's and Palmer's *A Natural History of Rape*. The authors argued that rape, even among humans, may be an adaptation. The disposition to rape is a genetically transmitted behavioural predisposition whose function is to enhance fitness, especially for males who, due to a combination of genetic and environmental circumstances, lack other ways to compete successfully for females. The authors' move from insects to humans is surely very problematic. We will return to this issue soon. However, here the authors' argument keeps to a level of generality that is supposed to make distinctions between species largely irrelevant. Still continuing the deductive argument:

12. For all males, under most circumstances, fitness tends to be enhanced by copulation at any time with any female.
13. For all females, under most circumstances, fitness tends to be enhanced by copulation only on some occasions with some males.
14. (*From 7 above*): Other things being equal, behavioural predispositions that enhance fitness will be perpetuated in the population.
15. (*From 12-14*): In the male population, the predisposition under most circumstances to copulate at any time with any female will be perpetuated, and in the female

---

21   Rossano, *Evolutionary Psychology*, 223-4.
22   Thornhill and Palmer, *A Natural History of Rape*, 53; Rossano, *Evolutionary Psychology*, 224-6.
23   Rossano, *Evolutionary Psychology*, 223-5.

population, the predisposition under most circumstances to copulate only on some occasions with some males will be perpetuated.

16. (*From 15*): Under certain environmental circumstances (limited access to predisposed females), males will be predisposed to copulate also with females who are not predisposed to copulate with them. This is defined as *forced copulation*.

17. (*From 14–16*): The predisposition under certain circumstances to use forced copulation will be perpetuated in the male population.

The argument appears again to make rationally grounded predictions about strivings and motives. However, it only speaks of predispositions. It says that, given the asymmetry between the optimal conditions for male and female reproductive success, male fitness is enhanced by mating also with females whose fitness is not enhanced by mating. But it does not claim that forced copulation is *used* by males *in order to* propagate their alleles.

Other authors have pointed out leaps and factual inaccuracies in Thornhill and Palmer's reasoning, especially as applied to the human case.[24] I will leave that on one side and focus on the alleged phenomenon of insect rape.

5. The identification of behaviour: Why 'rape'?

Thornhill and Palmer refer to earlier research, by Thornhill and others, on 'rape' among certain species of scorpionfly or *panorpidae*. Normally, the male *panorpa* that wants to mate attracts the female with a dead insect or nutritious salivary mass, then mounting the female while she is eating. But in some cases the male approaches the female empty-handed – for instance, because he has not found a dead insect or is not strong enough to retrieve it safely:

> The third option for a male is not to [offer] food at all, but to search actively for females. When a male of this sort encounters a prospective female, he darts towards her and grabs her by the wing or leg with his large genital forceps. . . . Although females appear to resist and often struggle free, occasionally a male is able to grasp the female's genitalia with his own and inseminate her. . . . There is no food transferred from the male to the female in these cases and the exercise has the appearance of forced copulation.[25]

---

[24] Karen Arnold, 'Evaluating Science on Epistemic and Moral Grounds (Formerly, Putting Anthropomorphism in Context),' Philsciarchive.pitt.edu/archive/00000373/00/Arnold.rtf (2000); Lisa E. Sanchez, 'How Homo Academicus Got His Name and Other Just-So Stories,' *Gender Issues*, 18 (2000), 83–103; Tony Ward and Richard Siegert, 'Rape and Evolutionary Psychology: A Critique of Thornhill and Palmer's Theory,' *Aggression and Violent Behaviour*, 7 (2002), 145–68.

[25] Randy Thornhill and John Alcock, *The Evolution of Insect Mating Systems* (Cambridge, MA: Harvard University Press, 1983), 272. See also Randy Thornhill, 'Rape in Panorpa Scorpionflies and a General Rape Hypothesis,' *Animal Behavior*, 28 (1980), 52–9, at 52, and Arnold, 'Evaluating Science on Epistemic and Moral Grounds,' 5.

Discussing this research, Karen Arnold points out several difficulties. If rape is defined as *forced* copulation, insemination or fertilization, the question arises what behaviour should count as a sign of coercion or consent among insects.[26] From Thornhill and Alcock it in fact emerges that an element of 'acceptance' seems to be involved:

> Females that have been grasped by a would-be "rapist" can probably refuse to permit the male ever to copulate with them by preventing entry of his aedeagus. But in some cases this response requires an extremely high degree of energy expenditure and, perhaps more important still, a considerable loss of time. . . . At some point his persistence may make it adaptive for the female to cut her losses and accept copulation in order to terminate the interaction with no further time and energy wasted. Such a mating would be a copulation of convenience, because the female gains no sperm-replenishment, material, or genetic benefits from the encounter.[27]

Arnold reports some writers arguing that the female's behaviour may just be an expression of unwillingness to be grasped in general.[28] Alternatively, it may be a case of the female testing the strength of potential mating partners, or 'coyness' to use the sort of anthropomorphic term typical of these discussions.[29] Thus *both* 'the requirement of gift' and 'escape attempt' could arguably be construed as the female's method to control the genetic quality of the inseminating male.

One solution by Thornhill is to define unwillingness regardless of questions about what insects *want*. Mating counts as 'rape' when it lies in the reproductive interest of the male but not of the female.[30] Mating is to the female insect's reproductive disadvantage when the male does not offer food. Similarly, in the quote above, the absence of any obvious benefit is given as the reason for describing the female's final acceptance of the male's sperm as only grudging.

Arnold points out that Thornhill is somewhat ambiguous. He does call rape *forced* copulation and describes the females as 'appear[ing] to resist'.[31] However, because of the difficulties of assessing insect motivation, his working definition of 'rape' must be formulated in terms of a clash of reproductive interests.[32] Arnold sums up:

> In summary, a behavior must meet the following three conditions in order for it to count as rape according to Thornhill: 1) the female is inseminated, or in the case of species with external fertilization, the female's eggs are fertilized,[[33]] 2) the female

26   Arnold, 'Evaluating Science on Epistemic and Moral Grounds,' 9–10.
27   Thornhill and Alcock, *The Evolution of Insect Mating Systems*, 469–70.
28   Arnold, 'Evaluating Science on Epistemic and Moral Grounds,' 9.
29   Ibid., 6. See also Thornhill and Alcock, *The Evolution of Insect Mating Systems*, 404.
30   Thornhill, 'Rape in Panorpa Scorpionflies and a General Rape Hypothesis,' 55–6; Arnold, 'Evaluating Science on Epistemic and Moral Grounds,' 7.
31   Thornhill and Alcock, *The Evolution of Insect Mating Systems*, 272.
32   Arnold, 'Evaluating Science on Epistemic and Moral Grounds,' 11, n. 2.
33   Thornhill, 'Rape in Panorpa Scorpionflies and a General Rape Hypothesis,' 52. However, Thornhill's later work suggests that copulation without insemination also may count as rape (Arnold, 'Evaluating Science on Epistemic and Moral Grounds,' 6–7, n. 1). Apparently, rape must, for Thornhill, at least involve some *probability* of insemination or fertilization (since rape could not otherwise count as a *reproductive* strategy).

is unwilling to be inseminated or have her eggs fertilized, where "unwilling" means that it is against her reproductive interests, 3) the male is willing to inseminate the female or fertilize her eggs, where "willing" means that it is in his reproductive interests to do so. On the basis of these criteria, Thornhill claims that male scorpionflies rape females.[34]

This approach has some important consequences, which Arnold spells out. Taken literally, it would imply that women past childbearing age cannot be raped, as no fertilization is even theoretically possible.[35] Furthermore, it means that we cannot tell whether rape has occurred from the psychological state of the female alone. We must also consider its impact on her fitness.[36] On the other hand, it is worth bearing in mind that, in the passage from Thornhill and Alcock quoted above, they place the term 'rapist' in scare quotes. They are themselves wary of using terms such as this without any caveat. Still, in order for the argument for human rape as a sexual adaptation to get through, Thornhill must assume a sufficient degree of continuity between human and insect cases of 'rape'.

As earlier with the praying mantis, there are two views here about how to define coercion and consent. One might think that the introduction of criteria for rape in terms of reproductive interests, instead of the female's visible resistance to insemination, is simply motivated by difficulties about entering the insect psyche. However, more is at issue: the question of where to look for explanations of conduct. Explaining animal behaviour on the basis of what the animals want goes against the main tendency of the explanatory framework proposed – the hope for 'ultimate' rather than 'proximate' explanations.

If the question simply is, 'Do insects rape?' it seems to me one might try to answer it empirically, independently of further theorizing. First, one would study insect behaviour and try to establish that some females are inseminated against their will. This might be difficult but unless we subscribe to complete agnosticism about other minds, there should be no a priori impossibility about, for example, finding that an insect avoids being mounted by a male.[37] We might be able to say, 'this insect wants to mate, this one doesn't, perhaps some don't care much'. This is indeed part of what Thornhill and others *have* done in their empirical research. Subsequently, one should consider how cases of insect 'rape' relate to the human crime for which the word 'rape' is used. The conclusion would doubtlessly be that both differences and similarities exist. Whether we wish to stress the former or the latter will depend on the context of the comparison, even though my guess is that, in most contexts, the differences will significantly outweigh the similarities. For instance, insect rape does not have

---

[34] Arnold, 'Evaluating Science on Epistemic and Moral Grounds,' 8.
[35] Ibid., 11.
[36] Ibid., n. 2.
[37] *Pace* Sanchez who completely rules out ascribing the comprehension of coercion and consent to insects (Sanchez, 'How Homo Academicus Got His Name and Other Just-So Stories,' 92). Arnold emphasizes that the task of determining unwillingness in animals is 'far from simple' ('Evaluating Science on Epistemic and Moral Grounds,' 6).

the ethical aspects of human rape. We will not condemn entomologists who fail to intervene. Not even animal rights activists have raised the issue.

In sum, I take it that there *is* unexplored space for research and informed speculation about what insects want or what they think, without immediately plunging into anthropomorphism. But formalizations of their 'reproductive strategies' are not contributions to *that* effort. Thornhill, Palmer and others may on the contrary be accused of muddying the waters for discussions of animal and human motivation. They certainly do not assume that the male who rapes literally does it in order to have young. Yet they maintain that rape is a *strategy*; in other words, that it represents a pattern at least implicitly presented as meaningful, rational and helpful for some purpose. My main point of criticism is that the quoted research hardly compares human beings and insects at all. It rather imposes a model of cost-benefit analysis on both.

6. Peter Winch on the identification of behaviour

A central question here is: How important is it to enter the minds of the individuals whose behaviour is being discussed? Unless we do so, what can we assert about what makes them act? Issues of this type have, for a long time, been discussed in the philosophy of the social sciences, especially in the context of cross-cultural comparisons. There the question is often framed in terms of a contrast between looking at behaviour from the participants' own point of view and looking at it from 'the outside'. It is lamentable, though quite understandable, that authors with a background in biology are not conversant with these debates.

As early as 1958, Peter Winch defended investigations that use the agents' own understanding of their behaviour as their starting point.[38] Winch was not just saying that, in the social sciences and humanities, the search for 'inside' understanding is *preferable* to an external perspective. His point was that no research into meaningful behaviour is even *possible* unless it somehow includes the agents' own understanding.[39] A theory that attempts to explain behaviour must first of all identify the thing to be explained. But when the 'thing' in question is meaningful behaviour, it is identified in terms that, implicitly or explicitly, presuppose references to how the agents would identify it.

Meaningful behaviour is, according to Winch, characterized by the fact that the relation between what agents do and the terms in which they would understand it

---

[38]   Peter Winch, *The Idea of a Social Science and Its Relation to Philosophy* (London: Routledge & Kegan Paul, 1958).

[39]   From an external point of view, behaviour that propagates copies of one's alleles might be construed as 'meaningful', just as travelling might be construed as a 'meaningful' – that is, efficient – way of spreading viruses across human populations. However, 'spreading alleles' and 'spreading viruses' are for the most part unintended side effects of some human behaviour whose intended meaning lies elsewhere. If one wants to know why humans spread viruses one needs to ask, for instance, why humans *travel*. The answers will be related to issues about motives and opportunities for travelling. There are other questions too, such as, 'What environmental factors make the spread of viruses possible?', but the answers will not be competing accounts of why humans travel.

is *internal*. They are not two separate things but two sides of a coin. '[O]bedience to a command' is a case in point.[40] The issuing of a command makes the subordinate perform what is commanded. But this is not just an instance of a stimulus causing a response. We must assume that both parties are aware that the 'stimulus' is a command and that they understand their own place in the hierarchy. Our justification for identifying the case as one of command and obedience (and not of physical coercion, reflex action, blackmail, etc.) lies in the fact that the agents see their mutual relation in that way. It would, for instance, not make sense to think that commands and obedience were in existence before human beings formed those concepts and started to apply them to behaviour.[41] Furthermore, it would be impossible to pin down some external, physical similarity of *behaviour* common to all cases of command and obedience. Thus Winch's criticism is directed against the idea that, for the sake of investigating the real causes of behaviour, one ought to identify a purely behavioural core that remains the same regardless of what the agents think.

Winch compared the situations of a natural and a social scientist. For both, their ability to frame meaningful general descriptions will depend on their employment of what Winch called 'judgements of identity'.[42] Researchers must be able to say, of two phenomena, that they fall under the same description – in other words, that they are in some respect instances of *the same* thing. But Winch argued that there is a difference between the natural and the social scientist. Identity judgements by the natural scientist will need to take account of prevailing theory and tradition within his field – say, nuclear physics. The social scientist, meanwhile, will investigate situations where agents have ideas about their own activities.[43]

> Two things may be called 'the same' or 'different' only with reference to a set of criteria which lay down what is to be regarded as a relevant difference. When the 'things' in question are purely physical the criteria appealed to will of course be those of the observer. But when one is dealing with intellectual (or, indeed, any kind of social) 'things', that is not so. For their *being* intellectual or social, as opposed to physical, in character depends entirely on their belonging in a certain way to a system of ideas or mode of living. . . . It is not open to [the social scientist] arbitrarily to impose his own standards from without.[44]

Identifying the agent's meaningful behaviour presupposes, according to Winch, already that one identifies concepts in terms of which the agent would see it. Conversely, by disregarding the agent's point of view one will also lose sight of the phenomenon to be studied. This is not to say that the social scientist may never go beyond the agent's 'unreflective kind of understanding' nor question it. But 'any more reflective

40 Winch, *The Idea of a Social Science and Its Relation to Philosophy*, 124.
41 Ibid., 125.
42 Ibid., 87.
43 Ibid., 86–8.
44 Ibid., 108.

understanding must necessarily presuppose, if it is to count as genuine understanding at all, the participant's unreflective understanding'.[45]

Winch would claim, for instance, that statements about motivation are not only tools that researchers use for explaining and predicting behaviour, by analogy with explanatory concepts in the natural sciences. The fact that human agents think of their actions in terms of *reasons* is *constitutive* of the identity of the actions in question. To say that someone is acting for a reason is to represent the action, at least in some sense, as reasonable from the point of view of the agent.[46]

In the discussion of human meaningful behaviour, the concepts of interest and strategy are internally connected to the idea of acting for a reason.[47] To say it is in a person's interest to do something is to claim that the person would have good reasons to do it. It would advance some goal that he or she would see as desirable. We certainly sometimes also say that people are blind to their true interest. But the intelligibility of that suggestion is dependent on the idea that, under the right circumstances, they could be reasonably brought over to see the value of the proposed goal. Thus to speak of what lies in someone's interest involves an attempt to address some issue from (what one takes to be, or what somehow ought to be) *that person's* point of view. To speak of the person as applying a strategy typically involves the further assumption that he or she consciously acts in order to bring about what lies in his or her interest. Sometimes, however, we may also speak of unconscious strategies.[48] Perhaps we should say: we can ascribe *unconscious* motives, interests and strategies to beings that it would make sense to imagine also having *conscious* motives, interests and strategies. What is unconscious might in principle be brought to the agent's consciousness.[49] In this sense, the assumption of 'inside perspective' is built into our normal use of these concepts.

In evolutionary explanations of behaviour, 'reproductive interest' and 'reproductive strategy' are clearly, and quite deliberately, not understood in the way that was just described. Scientists merely propose to look at unintended effects of behavioural patterns *as if* they followed from conscious intention. While this does not render these terms illegitimate here, it does raise the question how they should be understood. In this case, a suggestion of intelligible reasons for behaviour comes together with the admission that these are *not* reasons that the agents would see as reasonable or valuable, or could ever be brought to see as such. This simply means that reproductive interests and strategies *are not* interests and strategies in the sense in which these terms

45   Ibid., 89.
46   Ibid., 80–3.
47   On acting for a reason, see Winch, ibid., 46–8, 82.
48   Thus, as Nora Hämäläinen has pointed out to me, feminists often talk of 'strategies of power that (some) men use to limit the possibilities of women to act and talk in various situations at work, etc. These "strategies" are often unconscious, and yet [it may be] helpful to think of them as strategies'. Similarly, Marxists sometimes describe social phenomena in terms of class interests and class conflicts without implying that those involved think of their situation in that way. The crucial question here is, perhaps, how these descriptions should translate to a psychological level. Their use at least seems to presuppose a degree of rationality in the agents to whom they are applied.
49   See Winch's discussion of the ascription of unconscious motives in psychoanalysis in his *The Idea of a Social Science and Its Relation to Philosophy*, 89–90.

are normally applied to humans. Thus one thing that evolutionary psychology cannot do is to falsify our normal understanding of the motives of meaningful action, because it does not discuss motives (in the relevant sense) at all.

A second Winchian point concerns the criteria for identifying behaviour. Thornhill defined 'rape' in terms of the individuals' respective reproductive interests – thus hoping to side-step ethical and cultural issues, as well as questions about how to assess coercion in the animal world. Bearing Winch's argument in mind, one should ask whether insect rape, defined in this kind of abstract way, is sufficiently continuous with human rape to allow for meaningful comparisons.

Following Winch, an agent's behaviour can be described as a case of 'forcing someone to do something' only insofar as it is seen as (at least minimally) meaningful. It is seen as standing in a specific relation to the wishes and preferences of agents. Thus in the human case, one will need to clarify what constitutes sexual coercion in given societies. One must negotiate the murky borderlines between rape, seduction, prearranged marriage, sexual abuse, assault and many other activities. This even invites the question how helpful it is to call rape 'reproductive behaviour', implying parity with the raising of families. Most rapes neither result in, nor aim at, procreation.[50]

Furthermore, while copulation is technically a sexual act, this does not mean that the motivation for copulation (in rape or elsewhere) is always sexual. Humans, male or female, may have sex to please their partner, because they want to be loved, for money, for status reasons or in the service of some other need or goal.[51] Rape may sometimes be used as a tactic to retaliate or to control subordinates. As Tony Ward and Richard Siegert point out, 'rape, like any other behaviour, can express a number of goals and desires, the achievement of sexual pleasure is only one of these'.[52]

If the question simply is, 'why do males engage in copulation that is in their reproductive interest but not in that of the female?' it is not obvious that there is a problem to start with. The effects of a given instance of copulation on individual fitness will depend on various circumstances, many of which lie largely outside the individuals' control (genetic factors, the availability of mates, etc.). Thus asymmetries of this kind are simply to be expected. On the other hand, if the question is, 'why does the male scorpionfly approach the female without a gift?' the answer might be, 'it wants to mate but hasn't found a gift.' Finally, if the question is, 'why do rapes occur in human societies?' it is hard to see how one could say anything much without studying how sexual relations are perceived in the societies concerned, and how those perceptions interact with power relations, moral beliefs, family structure and similar factors. In other words, what one needs is qualitative social research.

[50]   Sanchez, 'How Homo Academicus Got His Name and Other Just-So Stories,' 87–8.
[51]   Ward and Siegert, 'Rape and Evolutionary Psychology: A Critique of Thornhill and Palmer's Theory,' 164.
[52]   Ibid.

7. Rationalism, anthropomorphism and the human animal

The facts of sexual reproduction and mortality belong to the predicament we share with our animal kin. Their importance in shaping human life projects and self-understanding can hardly be exaggerated. Winch, citing Giambattista Vico and T. S. Eliot, names 'birth, copulation and death' as the three significant events of human existence. They constitute 'limiting notions' which every culture somehow needs to accommodate: 'the very notion of human life is limited by these conceptions'.[53] The idea of limiting notions suggests that the opposition between cultural and biological explanations may be out of place here. In this case our biological predicament shapes culture, not because thinking is genetically determined, but because an *understanding* of this condition lies at the bottom of human purposeful agency. The importance of family and kinship in all societies is internally related to this. We are born into families and we expect our families to carry on life from where we leave it.

Rossano writes: 'It seems obvious that parents would expend considerable time, energy, and resources taking care of their young, because their offspring are their direct genetic contribution to the next generation.'[54] It does seem obvious, but one may ask exactly why it does so. Why is the next generation of any concern to us? According to evolutionary psychologists, it is because we are genetically programmed to propagate copies of our alleles. It seems to me, instead, that powerful cultural factors are at work here; factors that reflect the role of the Winchian limiting notions. The hope to see one's children reach adulthood is a very strong human motive. In part it is based on love, sometimes it involves considerations about a safe future, but often it also includes the somewhat self-centred element of wishing one's bloodline to continue. That may be called an example of someone literally adopting a reproductive *strategy*. On the other hand, many humans have no reproductive strategies at all in that sense.

Our familiarity with such hopes and pipedreams perhaps contributes to popular interest in evolutionary psychology. From a cultural point of view, the idea that we want to 'propagate our alleles' may appear plausible, and one might think there is no conflict between cultural and evolutionary explanations here. However, despite occasionally misleading popular work, the thesis of evolutionary psychologists is, in its bare essentials, very different from the traditional idea of purposefully continuing the family line.[55]

---

[53] Peter Winch, 'Understanding a Primitive Society', in his *Ethics and Action* (London: Routledge, 1972), 43–4.

[54] Rossano, *Evolutionary Psychology*, 17.

[55] It can be argued that indications of conscious planning among human populations to spread their alleles tend to weaken rather than strengthen the explanatory framework of evolutionary psychology. This is because conscious planning involves a cultural component, and the evolutionary mechanism is supposed to exist independently of culture.

Evolutionary psychologists sometimes talk as if animals were literally guided by complex strategies that may even lie beyond the grasp of many humans. However, the main problem with evolutionary psychology is not that of anthropomorphism. The problem is rather that it employs rationalist models that *seemingly* relate behaviour to motivation and rational planning. But the models are employed in a way that makes them independent of factual questions about motivation. Hence they cannot be claimed to reveal the hidden 'real' motivation behind human *or* animal action.[56]

[56] Thanks are due to Jan Antfolk, Mikel Burley, Niklas Forsberg and Nora Hämäläinen for careful and extremely helpful comments. This work is part of the research project *Westermarck and Beyond: Evolutionary Approaches to Morality and Their Critics*, financed by the Kone Foundation.

5

Three Perspectives on Altruism

Ylva Gustafsson

1. Introduction

The concept of altruism has long been one of vital importance in ethics. It is bound up with questions of human understanding and motivation as well as with questions concerning the form and origins of human social life. In this essay, I discuss three perspectives on altruism. I begin by describing Leda Cosmides' and John Tooby's thoughts on altruism. According to them, altruism consists of a mental reasoning mechanism by which we calculate how to act in order to maximize fitness when dealing with others. I contrast their view with Frans de Waal's thoughts on altruism as consisting of a mental mechanism of empathic imagination. Cosmides' and Tooby's as well as de Waal's perspectives are expressive of two problems: first, the reliance on an economic model of interpersonal relations; and secondly, the view of interpersonal understanding as consisting of an ability for analogical imagination. By reflecting on various examples of the natural and social form of our life I question the above mentioned perspectives and point to alternative ways of understanding altruism. In this context I also bring in Ludwig Wittgenstein's thoughts on pain, primitive reactions and language.

2. Tooby and Cosmides on altruism as a transaction

According to Cosmides and Tooby, the human mind has, over thousands of years, adapted so that we have become the social and moral beings we are today. Through evolutionary history, we have developed a mental skill of reasoning about how to act optimally in our dealings with others. All such dealings are formed so that they tend on the whole to enhance fitness (the propagation of copies of the agent's alleles in subsequent generations). Cosmides and Tooby describe this as a social contract theory:

> Social contract theory is based on the hypothesis that the human mind was designed by evolution to reliably develop a cognitive adaptation specialized for

reasoning about social exchange. . . . From an evolutionary point of view, the design of programs causing social behavior is constrained by the behavior of other agents. More precisely, it is constrained by the design of the behavior-regulating programs in other agents and the fitness consequences that result from the social interactions these programs cause.[1]

The fitness enhancing systems of reasoning to which our human minds have adapted, Cosmides and Tooby describe as 'evolutionarily stable strategies' or 'ESS'. According to them all our social engagement has evolved because it has proved to be an ESS. If our social engagement with others would not enhance fitness it could not have survived for thousands of years, is their thought.

> An *evolutionarily stable strategy* (ESS) is a strategy (a decision rule) that can persist in a population because it produces fitness outcomes greater than or equal to alternative strategies (Maynard Smith, 1982). The rules of reasoning and decision making that guide social exchange in humans would not exist unless they had outcompeted alternatives, so we should expect that they implement an ESS.[2]

In another article, Tooby and Cosmides et al. conclude that altruism is the outcome of a non-conscious mental mechanism whereby we calculate how to cooperate with others.[3] They note that human beings tend to be more altruistic towards their family than towards other people. This they define as 'kin selection'. Kin selection is, according to them, an ESS that enhances fitness. Tooby et al. also speak in this context of a 'welfare trade-off ratio' or 'WTR'. By this they mean a mental mechanism or a 'variable' that regulates how much I ought to help others in order to gain certain benefits for myself. This variable can be 'upregulated' when we are dealing with genetic relatives.

> [N]atural selection should have designed the human motivational architecture to embody programs determining how high one's welfare trade-off ratio toward other individuals should be set. . . . kin selection theory tells us that, all else equal, WTR should be upregulated for close genetic relatives, motivating us to help kin more and harm them less than we otherwise would.[4]

The degree to which we make 'sacrifices' for others is dependent on a 'kinship index' or 'kin selection'. People also use a 'WTR' index in order to calculate how much a

---

[1] Leda Cosmides and John Tooby, 'Can a General Deontic Logic Capture the Facts of Human Moral Reasoning? How the Mind Interprets Social Exchange Rules and Detects Cheaters,' in *Moral Psychology, Vol. 1: The Evolution of Morality: Adaptations and Innateness*, ed. Walter Sinnott-Armstrong (Cambridge, MA: MIT Press, 2008), 69–70.

[2] Cosmides and Tooby, 'Can a General Deontic Logic Capture the Facts of Human Moral Reasoning?,' 70. The work that they cite by Maynard Smith is: J. Maynard Smith, *Evolution and the Theory of Games* (Cambridge: Cambridge University Press, 1982).

[3] John Tooby et al., 'Internal Regulatory Variables and the Design of Human Motivation: A Computational and Evolutionary Approach,' in *Handbook of Approach and Avoidance Motivation*, ed. Andrew Elliot (Mahwah, NJ: Lawrence Erlbaum Associates, 2008).

[4] Ibid., 260.

particular person is willing to sacrifice his or her own welfare for yours. Important to note here is also that Tooby et al. think of these systems of reasoning as occurring non-consciously.

> [T]he welfare trade-off ratio, WTR is an internal regulatory variable expressing how much you value j's welfare relative to your own. Its value is nonconsciously expressed in many decisions you make throughout the day – how much chocolate you leave for j, how loud you play your music when j is trying to work, whether to clean up the mess or leave it for j, whether to call home to let j know you will be late. It is computed by a system, the welfare trade-off ratio estimator, that takes into account a specific array of relevant variables. . . .[5]

Cosmides and Tooby follow a pattern of thinking that can be traced to Thomas Hobbes in thinking that human social life can be described in economical terms grounded on a social contract.[6] However, for Hobbes the social contract was a theoretical explanation of the origins of society, but not centrally an internalistic mentalistic one nor a mathematical biological one. The linkage between a mathematical, economical and biological perspective comes later, with William D. Hamilton and John Maynard Smith. Hamilton argued that an organism's survival ought to be understood according to 'inclusive fitness'. By inclusive fitness Hamilton meant that an organism's genes have higher chances to survive if the organism cooperates with relatives or with other organisms of the same species. This Hamilton thought to be an explaining factor for why certain species are altruistic. Hamilton writes for instance:

> The social behaviour of a species evolves in such a way that in each distinct behaviour-evoking situation the individual will seem to value his neighbour's fitness against his own according to the coefficients of relationship appropriate to that situation.[7]

Following Hamilton's thoughts on inclusive fitness, John Maynard Smith introduced the thought of 'ESS' and argued that 'kin selection' and 'group selection' are such basic ESS. J. M. Smith also introduced the economical principle of game theory to his reflections on evolution. Cosmides' and Tooby's thinking largely follow this line of thought.

3. Frans de Waal and Adam Smith on our natural social being

Frans de Waal's perspective on social life and on altruism differs from Cosmides' and Tooby's perspective. De Waal is sceptical of a transactional perspective where

---

[5] Ibid., 261.

[6] See, for instance, Hobbes, *Leviathan* (Cambridge: Cambridge University Press, 1996), chs. 13–15.

[7] William D. Hamilton, 'The Genetical Evolution of Social Behaviour: II,' *Journal of Theoretical Biology*, 7 (1964), 19.

we fulfil other people's wishes in order to gain personal advantages. He sees altruism as spontaneous emotional responsiveness that evolves naturally as we grow up with others. 'Since expressions of sympathy emerge at an early age in virtually every member of our species, they are as natural as the first step.'[8] De Waal sees ape and human life as a shared life where the fact that others care for us is a natural aspect of how we grow up and develop emotional responsiveness to these others. He considers apes and human beings as born into interpersonal relations and, in this regard, also as social from the very start.

De Waal's perspective can be seen as linked with a different tradition than Cosmides' and Tooby's. De Waal follows a tradition of thought that can be traced back to Adam Smith's philosophy.[9] According to Smith, human beings have a natural inclination to form social bonds and relations and to feel compassion for others.

> How selfish soever man may be supposed, there are evidently some principles in his nature, which interest him in the fortune of others, and render their happiness necessary to him, though he derives nothing from it except the pleasure of seeing it. Of this kind is pity or compassion, the emotion which we feel for the misery of others, when we either see it, or are made to conceive it in a very lively manner. That we often derive sorrow from the sorrows of others, is a matter of fact too obvious to require any instances to prove it; for this sentiment, like all the other original passions of human nature, is by no means confined to the virtuous and humane, though they perhaps may feel it with the most exquisite sensibility. The greatest ruffian, the most hardened violator of the laws of society, is not altogether without it.[10]

Smith thinks we spontaneously care for others *without* requiring anything in return.[11] We are simply affected by others. Think again of Tooby's and Cosmides' description of such acts as calling home to let one's family know one will be late. Instead of describing this as a 'trade off' response, one could say that calling home to let one's family know one will be late is an expression of spontaneous considerateness and care. It is usually not a

---

8   Frans de Waal, 'Morally Evolved, Primate Social Instincts, Human Morality and the Rise and Fall of "Veneer Theory"', in *Primates and Philosophers: How Morality Evolved*, ed. Stephen Macedo and Josiah Ober (Princeton, NJ: Princeton University Press, 2006), 28.

9   De Waal's perspective on altruism also has similarities with Edward Westermarck's reciprocal conception of emotions. Westermarck writes: 'The moral emotions are retributive emotions. A retributive emotion is a reactive attitude of mind, either hostile or kindly, towards a living being (or something taken for a living being), regarded as a cause of pain or pleasure' (*Ethical Relativity* (London: Kegan Paul, Trench, Trubner & Co., 1932), 172).

10   Adam Smith, *The Theory of Moral Sentiments*, ed. Knud Haakonsen (Cambridge: Cambridge University Press, 2002), 11.

11   Smith does not specifically use the word 'altruism', as there was no such word at the time when he was writing. The French term *altruisme* was first introduced into philosophical theory by Auguste Comte in his *Catéchisme Positiviste* (Paris: Carilian-Goeury and Vor Dalmont, 1852), and this term was translated as 'altruism' in Richard Congreve's English rendering of the text (*The Catechism of Positive Religion*, trans. Richard Congreve (London: Chapman, 1858)). Smith's thoughts on compassion are, however, similar to Comte's perspective on altruism, inasmuch as Comte does not think of altruism as being based on selfish purposes.

decision we make on the basis of a calculation, nor is it because we expect something in return that we call home. Very often it is simply something we do because otherwise the family would get worried. It is a response that in itself is expressive of the close relation.

However, it is not evident that Cosmides and Tooby would feel that Smith's and de Waal's thoughts constitute a counterargument. When de Waal and Smith both state that our compassionate responses are not based on our trying to gain benefits, Cosmides and Tooby might reply by saying that of course we often respond compassionately in ways that do not *appear* to involve calculation; it is simply the case that these calculations occur *non-consciously*. I shall return to reflect on Cosmides and Tooby later in this essay.

4. Sympathy, empathy and analogical imagination

Instead of explaining our care for others as based on a non-conscious reasoning mechanism for fitness enhancement, Adam Smith sees compassion as connected with our capacity to imagine ourselves in the other's situation. This shows in his thoughts on sympathy. By sympathy, Smith does not mean compassion but any kind of imaginative experience of another person's feelings that cause us to be emotionally moved by the other. This is, in modern terms, called empathy. The explanation for our spontaneous reactions of compassion must be sought in the function of our mind and in our capacity to imagine how others feel.

> As we have no immediate experience of what other men feel, we can form no idea of the manner in which they are affected, but by conceiving what we ourselves should feel in the like situation. Though our brother is upon the rack, as long as we ourselves are at our ease, our senses will never inform us of what he suffers. They never did, and never can, carry us beyond our own person, and it is by the imagination only that we can form any conception of what are his sensations. Neither can that faculty help us to this any other way, than by representing to us what would be our own, if we were in his case. It is the impressions of our own senses only, not those of his, which our imaginations copy. By the imagination we place ourselves in his situation, we conceive ourselves enduring all the same torments, we enter as it were into his body, and become in some measure the same person with him, and thence form some idea of his sensations, and even feel something which, though weaker in degree, is not altogether unlike them.[12]

Smith's explanation of moral sentiments consists in an argument from analogy. Sympathy is simply the effect of our capacity to imagine ourselves in the other's shoes. This effect can in turn make us feel compassion for the other person's misery.

---

[12]   Smith, *The Theory of Moral Sentiments*, 11–12.

De Waal thinks along similar lines, that human beings and apes have the capacity to imagine how other people (or apes) feel. He uses, however, the word 'empathy' instead of the word 'sympathy'. According to de Waal our capacity to imagine is something that develops gradually. First we simply have a tendency for what he calls 'emotional linkage' or 'emotional contagion'. Eventually we also develop the capacity to see that others have separate perspectives on reality. This is part of what it means to feel empathy.

> When the emotional state of one individual induces a matching or closely related state in another, we speak of "emotional contagion". . . . With increasing differentiation between self and other, and an increasing appreciation of the precise circumstances underlying the emotional states of others, emotional contagion develops into empathy. Empathy encompasses – and could not possibly have arisen without – emotional contagion, but it goes beyond it in that it places filters between the other's and one's own state.[13]

This capacity for empathy is, according to de Waal, not something that can be explained as deriving from social competition, but derives from a need for cooperation. Human beings as well as apes have a natural inclination to become emotionally affected by others because we have a natural need to cooperate.

> I am personally convinced that apes take one another's perspective, and that the evolutionary origin of this ability is not to be sought in social competition, even if it is readily applied in this domain (Hare and Tomasello 2004), but in the need for cooperation. At the core of perspective-taking is emotional linkage between individuals – widespread in social mammals – upon which evolution (or development) builds ever more complex manifestations, including appraisal of another's knowledge and intentions.[14]

This 'emotional linkage' leads eventually to a more advanced capacity to imagine another's perspective, something de Waal calls 'empathy'. De Waal thinks interpersonal understanding originates from an analogical bodily responsiveness (or imitation) that eventually leads to analogical imagination. This analogical ability to imagine enables us to feel compassion for others.

> [A]t the core of the empathic capacity is a relatively simple mechanism that provides an observer (the "subject") with access to the emotional state of another (the "object") through the subject's own neural and bodily representations. When the subject attends to the object's state, the subject's neural representations of similar states are automatically activated. The closer and more similar subject

---

[13] De Waal, *Primates and Philosophers: How Morality Evolved*, 26.

[14] Ibid., 72. The work that de Waal cites by Hare and Tomasello is: Brian Hare and Michael Tomasello, 'Chimpanzees are More Skilful in Competitive than in Cooperative Cognitive Tasks,' *Animal Behaviour* 68 (2004), 571–81.

and object are, the easier it will be for the subject's perception to activate motor and autonomic responses that match the object's (e.g., changes in heart rate, skin conductance, facial expression, body posture). This activation allows the subject to "get under the skin" of the object, sharing its feelings and needs, which embodiment in turn fosters sympathy, compassion, and helping.[15]

I agree with de Waal that it is a common feature that we are emotionally affected by others and that a kind of emotional contagion often occurs. But the question is whether this can be understood as expressive of an analogical mental mechanism, as de Waal describes it. And the question is also whether compassion can be understood in this sense as deriving from analogical imagination.

To begin with, let us reflect a bit on an example of emotional contagion. Think about what it can mean to play some simple game with a baby such as peek-a-boo. One could say that the child is affected by the parent's joy and that emotional contagion occurs. Likewise one could say that the parent is affected by the child's joy. But on the other hand there is something more to this. They are in the game of peek-a-boo *sharing* a moment of joy. From de Waal's description of emotional contagion it appears that one person is emotionally affected by the other but that the feeling itself is something individual. It appears that there are two people who have the same private feelings but that the shared form of playing, that we play *together*, is not a part of what the experience means. In reality, however, when we play a game together it is this playing *together* that is fun. It would not be the same feeling of joy if the child was playing alone. De Waal's perspective on emotional contagion is problematic because he suggests all our experiences basically are inner, private phenomena. This makes him unable to really see what it means to share an experience with another.

How is it then with de Waal's thought that compassion derives from analogical imagination? The idea that compassion is based on the use of an analogical method of imagination is expressive of a tendency not to see the difference between a first person perspective and a second person perspective when we talk about sensations. The thought is that the second person perspective on sensations is a kind of copy of the first person perspective. Ludwig Wittgenstein is critical of this way of thinking. In *Zettel* he writes:

> "Putting the cart before the horse" may be said of an explanation like the following: we tend someone else because by analogy with our own case we believe that he is experiencing pain too. – Instead of saying: Get to know a new aspect from this special chapter of human behaviour – from this use of language.[16]

A bit earlier he also writes:

> It is a help here to remember that it is a primitive reaction to tend, to treat, the part that hurts when someone else is in pain; and not merely when oneself is – and so

[15] De Waal, *Primates and Philosophers: How Morality Evolved*, 37–8.
[16] Ludwig Wittgenstein, *Zettel* (Oxford: Blackwell, 1967), §542.

to pay attention to other people's pain-behaviour, as one does *not* pay attention to one's own pain behaviour.[17]

Wittgenstein does not think of compassion as something based on our ability to imagine having similar sensations of pain as the other. On the contrary, he talks of how our reaction to attend to other people's pain is just as basic a feature in our life as attending to one's own pain; it does not derive from analogical imagination. He also points out that we do not pay attention to our own behaviour when we are in pain as we do when others are in pain. Wittgenstein is here also criticizing the thought that the word 'pain' would refer to an inner sensation that we from a first-person point of view have privileged access to. The argument from analogy builds on such a thought of privileged access to our own inner states while we do not have such access to other people's inner states. It also builds on an epistemological conception of knowledge and understanding where knowing something means to have information about something rather than *doing* something. Instead of thinking of concepts like imagination and attention as basically epistemological concepts Wittgenstein shows them to be concepts that gain their meaning in our *responding* to the other. This also means that these concepts have a moral form. In *Philosophical Investigations* Wittgenstein writes: 'How am I filled with pity *for this man*? How does it come out what the object of my pity is? (Pity, one may say, is a form of conviction that someone else is in pain.)'[18]

In *Zettel* he writes: 'Only surrounded by certain normal manifestations of life, is there such a thing as an expression of pain. Only surrounded by an even more far-reaching particular manifestation of life, such a thing as the expression of sorrow or affection. And so on.'[19] One could say that de Waal and Wittgenstein have different perspectives on what counts as 'natural' in human life. For de Waal it is important that our bodies resemble each other and that our minds work in similar ways. This he thinks enables us to understand each other and respond to each other. For Wittgenstein on the other hand it is with the background of a broad shared pattern of life that the concept of pain has meaning – a life where we naturally share close relations with others and where we then also respond to each other in various ways, by for instance displaying sorrow or affection. My aim here has not been to say that we *never* have experiences we might think of as private, but to emphasize the fact that often our experiences are *shared* in the sense that we *do something together*. De Waal's attempt to describe such experiences as something inner occurring simultaneously in two people's minds changes the meaning of such experiences into something private. It is only in specific life contexts that the thought of certain feelings as private has meaning, and it is only in specific life contexts that it can sometimes be meaningful to think of people as being separate. Such a context might for instance be when a person feels ashamed over something and therefore feels he cannot talk with others. Sometimes we can also feel helpless in front of another person's suffering, not being able to reach her or to comfort her. It can be an experience of separateness. Such an

---

[17] Ibid., §540.
[18] Ludwig Wittgenstein, *Philosophical Investigations*, 3rd edn (Oxford: Blackwell, 2001), §287.
[19] Wittgenstein, *Zettel*, §534.

experience of separateness is, however, in itself a moral reaction towards the other's suffering.

The fact that we usually talk with each other when we want to understand each other also shows how our knowledge and understanding take moral and relational form. Think about the difference between finding things out about another person's feelings behind her back (for instance by secretly reading a person's diary) and, on the other hand, *asking* her what she feels. Even if it might be the case that we sometimes get to know more about a person by finding things out about her behind her back, there is an important moral side in the fact that we *ask* the other person what she thinks or wants. Asking is connected with honesty, respect and closeness. The wish to talk with the other and hear *her* words in telling things is often in itself both an expression of moral acknowledgement of the other and also a wish to share her life. Another person's perspective has meaning here in that we really want to hear her words. But her having a separate perspective also has meaning in the sense that I respect the fact that she might *not* always want to talk with me; I respect her wish to keep her feelings to herself. In this sense the concept of understanding gets its form through the moral and personal relationship that we stand in with the other, where our telling each other things, asking each other things, is central for the character of our relationship. This cannot be seen from an epistemological point of view where understanding is thought of as merely a matter of gaining as much information as possible about what goes on inside the other's mind. David Hamlyn writes:

> Complete or full understanding of a person is impossible without standing in a personal relationship to him (and to say this is not *merely* to say that without this I should not have sufficient information about him, if that information is construed in an impersonal way, nor is it to say that without this I should not have the tacit knowledge how to cope with him and things of that kind). . . . Understanding of people is impossible altogether without knowledge of what a person is, and this implies a foundation in what Wittgenstein would have called features of our form of life, and what I have referred to as 'natural reactions of person to person'. It is on this basis, as I see it, that interpersonal understanding develops, and without it it would never get off the ground.[20]

Even if de Waal and Adam Smith often have a good sense for the social character of human life, they are caught in a dualistic and epistemological picture of human understanding. This makes them unable to really see what it can mean to share a human life with others, to share an experience with another, to respond to another's suffering, to talk with another, to grow into a close relationship with another or to feel separate from another.

[20] D. W. Hamlyn, *Perception, Learning and the Self: Essays in the Philosophy of Psychology* (Hampshire: Gregg Revivals, 1983), 237.

5. Eating

Connected with de Waal's thoughts on altruism are his thoughts on cooperation. J. C. Flack and de Waal describe food sharing among chimpanzees:

> Food sharing is known in chimpanzees. . . . It is an alternative method to social dominance and direct competition by which adult members of a social group distribute resources among themselves. Most food sharing requires fine-tuned communication about intentions and desires in order to facilitate inter-individual food transfers.[21]

They further explain this food sharing: '[T]he reciprocity hypothesis . . . proposes that food sharing is part of a system of mutual obligations that can involve material exchange, the exchange of social favours such as grooming and agonistic support, or some combination of the two.'[22] Flack's and de Waal's thesis is that animal social engagement can be understood as patterns of reciprocal emotional responsiveness useful for the group and that those patterns have evolved for that reason. They also see such reciprocal responsiveness as a basic pattern in altruism. The basic factors that unite human beings have to do with, on the one hand, the fact that we are able to imagine how others feel and, on the other hand, our need to cooperate, for instance by sharing food. Flack's and de Waal's conception of cooperation is reminiscent here of Cosmides' and Tooby's transactional perspective; exchanges of favours are efforts to fulfil another person's desires, whatever these desires may be. And they are carried out in order to secure certain benefits.

But leaving that aspect aside, there is another feature of the account offered by Flack and de Waal that I want to reflect on some more, namely how they describe eating. Think about how Flack and de Waal talk about 'food sharing'. Is it a good description of what it means for human beings to eat together, to describe this as a method of social exchange of favours? And is this the *only* or *most basic* description of what it means to eat together? When human beings eat together it is often a way of spending time together, *being together*. We do not simply exchange food in order to cooperate; we eat *together*. That is, for human beings eating cannot be described as mere exchanges of favours. Eating *together* has a meaning that is different from that which eating alone has. I agree with Flack and de Waal that eating together is a natural feature of our life and that sharing food is of practical importance. But their portrayal of this as an exchange of favours is problematic. In fact, our eating together is often expressive of how our lives are intimately woven together in multifarious ways. There is a sense in which we share a meal with our children, where this is so self-evident that it cannot be understood as cooperation or as an exchange of benefits. Sharing meals

---

[21]   Jessica C. Flack and Frans de Waal, ' "Any Animal Whatever": Darwinian Building Blocks of Morality in Monkeys and Apes,' in *Evolutionary Origins of Morality: Cross Disciplinary Perspectives*, ed. Leonard D. Katz (Thorverton: Imprint Academic, 2000), 4.

[22]   Ibid., 5.

brings meaning to the whole situation. At the same time, the fact that we share meals is expressive of our sense of responsibility for the other. This is a responsibility that cannot be understood without acknowledging the importance of our shared presence for each other in eating together. This presence shows itself, for instance, in the fact that we often spontaneously talk with each other and listen to each other while eating.

According to Flack and de Waal communication is used in food sharing in order to 'facilitate inter-individual food transfers'.[23] I agree that talking while eating can be of practical importance, such as me asking you to pass the salt. However, the fact that we often talk while we eat also often has a non-practical character. We simply talk because we are together. Our spontaneous readiness to talk with each other is an expression of our presence for the other. In this sense there is something similar in how we can be drawn into playing peek-a-boo with a child and in how we are often spontaneously drawn into conversations while eating with others. This may be something neither of us *decides* to do; and it has no further purpose. But it is expressive of how we can be there for each other. This form of spontaneous presence is in itself expressive of a moral sensitivity towards the other; an ability to be drawn into her life, to take her life seriously.

Of course there can be situations where people do not talk with each other while eating. In some families it might be even more of a rule that they do not talk while eating. There can be various reasons for this. One reason can be that the married couple are deeply fed up of each other. In such situations a cold manner of talking merely in order to 'facilitate inter-individual food transfers' might also take place. But my point above has been to say that even if both animals and human beings might sometimes share meals in a transactional manner, and even if both animals and human beings sometimes communicate simply in order to exchange benefits while eating, these are not any more basic or more natural ways of eating than the ways we can enjoy eating *together*. Nor do these transactional ways of eating and talking explain the multifarious ways we share a life with each other and find meaning in being together.

Wittgenstein writes: 'Commanding, questioning, storytelling, chatting, are as much a part of our natural history as walking, eating, drinking, playing.'[24] One thing Wittgenstein suggests here is that it is problematic to think of human nature as if the natural part of it only consisted in our doing practical things in order to survive, while language is thought of as a 'cultural' feature extraneous to it. Wittgenstein sees talking as a feature of our lives that is no less natural than eating. But it is not only that eating and talking both are natural, as if these still were separate aspects of life. What it means for human beings to eat cannot be separated from the fact that we share meals and that we commonly spontaneously talk while we eat. In this sense our physical needs largely get their form and meaning from the fact that we live a shared life with others, and a shared life full of talk. Another thing he suggests is that it is problematic to think of language as basically a practical tool. Wittgenstein mentions such things as chatting, storytelling and playing, that is, ways of talking and doing things that do not have a

---

specific goal but are rather forms of being with others. For Wittgenstein, our practical actions are not in any sense more basic or more natural than our non-practical ones. Nor is a practical use of language, such as exchanging information, more basic than a non-practical one such as chatting, or joking, or storytelling.

In his later work, *The Age of Empathy* de Waal seems, however, to acknowledge the fact that eating cannot simply be described as an exchange of favours. There he notes that apes not only help each other out when trying to get food but they also sit down and eat together. He asks:

> . . . could it be that they just love to eat together? If both monkeys are rewarded, they will sit side by side munching on the same food. Do things taste better together than alone, the way we are more at ease having dinner with family and friends?[25]

De Waal then draws the following conclusion:

> Perhaps it is time to abandon the idea that individuals faced with others in need decide whether to help, or not, by mentally tallying up costs and benefits. These calculations have likely been made for them by natural selection. Weighing the consequences of behavior over evolutionary time, it has endowed primates with empathy, which ensures that they help others under the right circumstances. The fact that empathy is most easily aroused by familiar partners guarantees that assistance flows chiefly toward those close to the actor.[26]

Despite the fact that de Waal notes that apes (and humans) enjoy eating together, he is unable to really see what it means to grow into a close personal relation, and how our eating together is part of this. For him our enjoying eating together is expressive of empathy which in turn is useful for survival since it makes us assist the ones that are close to us. I have earlier in this essay questioned de Waal's thoughts on empathy. I have also above tried to show that the social elements of our lives cannot be understood simply as consequences of the needs for practical cooperation. Our practical help and care are, on the contrary, an integral part of a larger pattern of acknowledging each other as persons. What it means to understand a child's physical needs cannot be separated from our acknowledging this child as someone to be with also in other ways, someone to eat together with, someone to chat with, quarrel with, and tell stories to, someone with whom we share a future life and a history. This relation cannot be understood in terms of empathic analogical imagination, nor can our social relations be well understood as motivated exclusively by our physical or biological needs.

> The child becomes progressively initiated into forms of life that those around it engage in and follow out continually; and this is possible because the child, as

---

[25] Frans de Waal, *The Age of Empathy: Nature's Lessons for a Kinder Society* (New York: Harmony, 2009), 113.
[26] Ibid., 115.

a result of its natural constitution, is a potential sharer in these forms of life and needs to be made an actual sharer in them by being constantly treated as one. In a word, the child comes to communicate by being communicated with.[27]

6. The idea of choice

I shall now return to reflect once more on Cosmides' and Tooby's idea of altruism as transactions. Two features are central in their transactional perspective. First, human interpersonal encounters are described as if these were always based on individual preferences. That we engage in social interaction is something we do because it is useful for our own survival and fitness. Second, Cosmides and Tooby also describe our transactional reasoning as taking place on a non-concious level. This unconscious level of reasoning they think of as a kind of mathematical reasoning function that governs all our social engagement. All social engagement is thought of as taking the form of exchanges of benefits. But how are these two features to be understood? Cosmides and Tooby defend their view of a non-conscious mental transactional reasoning function by referring to certain empirical evidence or examples from ordinary life. Some such examples were mentioned near the beginning of this essay: 'how much chocolate you leave for j, how loud you play your music when j is trying to work, whether to clean up the mess or leave it for j. . . .' Surely we do often think along these lines, but does this prove that there is an underlying non-conscious mental function of fitness reasoning taking place? It seems to me that these are only a bunch of examples described in a one-sided manner. However, Cosmides and Tooby also refer to certain empirical observations of 'primitive' people. With the example of how a 'primitive' !Kung San woman reasons about food exchanges they want to point to the natural as well as universal forms of economical reasoning in social situations.

> When Agent X provides a benefit to Agent Y, triggering the expectation in both that Y will at some point provide a benefit to X in return, a social exchange relationship has been initiated. Indeed, within hunter-gatherer bands, many or most reciprocity interactions are implicit.[28]

To illustrate their point, Cosmides and Tooby then offer the following quotation from Nisa, a !Kung San gatherer from Botswana who was interviewed by Marjorie Shostak:

> If a person doesn't give something to me, I won't give anything to that person. If I'm sitting eating, and someone like that comes by, I say, "Uhn, uhn. I'm not going to give any of this to you. When you have food, the things you do with it make me

---

[27] Hamlyn, *Perception, Learning and the Self: Essays in the Philosophy of Psychology*, 106.

[28] Cosmides and Tooby, 'Can a General Deontic Logic Capture the Facts of Human Moral Reasoning? How the Mind Interprets Social Exchange Rules and Detects Cheaters', 72.

unhappy. If you even once in a while gave me something nice, I would surely give some of this to you."[29]

Commenting on this passage, Cosmides and Tooby then add:

> Nisa's words express her expectations about social exchange, which form an implicit social contract: If you are to get food in the future from me, then you must share food with me. Whether we are San foragers or city dwellers, we all realize that the act of accepting a benefit from someone triggers an obligation to behave in a way that somehow benefits the provider, now or in the future.[30]

I do not want to deny that these ways of thinking of food sharing might be found anywhere among human beings, but the question is whether the example proves that there is an underlying non-conscious economical pattern of reasoning in all social engagement. Cosmides and Tooby seem to think that anything that is done among hunter gatherers must prove the feature to be a basic natural feature of all human reasoning. My impression is, however, that the woman quoted above is bitter about others, just as anyone can become bitter. This is a woman who has become fed up with sharing her food with others because others have sometimes not shared their food with her. But that a person sometimes grows bitter does not prove that this is how we all non-consciously 'work' socially. Her comment is not proof of some general, natural, underlying social pattern of reasoning merely because she lives the life of a hunter gatherer. Nor is the fact that all people around the world *sometimes* become bitter proof that social life has a basically transactional function. There are also a lot of people in the world who are not bitter, and probably there are also !Kung San people who are not bitter all the time. Besides, when a person becomes bitter we usually do not say that she behaves naturally. But we might say that she has become sick in her heart. Cosmides and Tooby think the example of a 'primitive' person's reasoning proves the general and natural transactional form underlying all human social engagement. However, this way of taking 'primitive' people as proof of some basic underlying pattern of thinking is problematic.

Cosmides' and Tooby's non-conscious transactional perspective is also connected with a predilection for an economic jargon when describing human relations. So they talk about 'kin selection', 'welfare trade-off ratio', etc. when describing our care for family members. These expressions are thought of as describing general underlying behavioural patterns. Our ordinary relational words such as 'child' or 'parent' are thought to gain their meaning from these underlying economical behavioural patterns. These economical expressions build, as I mentioned earlier, on a conception of human relations as based on individual preferences. The idea is that we can choose to take any kind of attitude towards

[29] Marjorie Shostak, *Nisa: The Life and Words of a !Kung Woman* (Cambridge, MA: Harvard University Press, 1981), 89; quoted in Cosmides and Tooby, 'Can a General Deontic Logic Capture the Facts of Human Moral Reasoning? How the Mind Interprets Social Exchange Rules and Detects Cheaters,' 72.

[30] Cosmides and Tooby, 'Can a General Deontic Logic Capture the Facts of Human Moral Reasoning? How the Mind Interprets Social Exchange Rules and Detects Cheaters,' 72.

another person's life, provided this attitude benefits our own chances of survival. Other people's lives are in this sense secondary to my own life. However, Knud Løgstrup writes: 'The other person is in such a real sense a part of our world that it is in fact awkward to refer to him or her as "the other person" rather than as one's child or spouse.'[31] As Løgstrup notes, the words 'child' and 'spouse' do not mean 'second person'. In this sense there is a big conceptual difference between Cosmides' and Tooby's use of expressions such as 'kin selection' and on the other hand Løgstrup using the word 'child'. The word 'child' implies a human relation where the child is involved in a certain way of living with others, sharing days and years in close relations, and where there are people who talk with and care for this child. These ways of sharing a life are internal to the self-evident form our responsibility for the child takes. From a perspective where all interpersonal relations are described as based on personal preferences it will not be possible to understand what it means to grow into a close relationship with another. Nor can our responsibility for one another be understood from such a perspective.

Think also about what it means to say that a person is dying. This is not simply a neutral description of the person's physical state: it is a moral description. When we say that a person is dying we also usually say that there is nothing more we can do to help him survive. At the same time, that a person is dying and that we cannot help him to survive usually does not mean that we leave him. The awareness of another person's coming death often makes us attend to him in a special way. In particular, it typically prompts the dying person's close ones to attend to him. We try to make his last days as bearable as possible; we try to ease possible pain, we help practically, but we also talk, we share meals together, and we often share thoughts and memories about life and about loved ones. By this I do not mean that we necessarily always help our close ones or other people who are dying. Family relations can sometimes be deeply injured, filled with years of quarrel that tear people apart. People can also wish for the death of another and kill others. But these attitudes are in no sense more basic than the ways we often do care for others.

I have argued three things here. First, words like 'child' or 'parent' cannot be understood from a perspective where interpersonal relations are thought of merely as transactions. Secondly, words such as 'birth', 'illness' and 'death' largely get their meaning from our standing in close personal relations to others. Thirdly, our understanding another person's physical needs is largely integral to our acknowledging the person also in other ways, such as our talking with her and listening to her. Conceptions of altruism that do not acknowledge these three aspects will not be able to show the natural form of our shared human life.[32]

[31] Knud Ejler Løgstrup, *The Ethical Demand* (Notre Dame, IN: University of Notre Dame Press, 1994), 125.

[32] The work with this essay has been funded by the Kone Foundation project *Westermarck and Beyond: Evolutionary Approaches to Morality and Their Critics*, as well as by the Finnish Academy of Science project *Emotions in Dialogue: Perspectives from the Humanitites*. I am grateful for valuable help and comments from Mikel Burley, Niklas Forsberg, Lars Hertzberg, Nora Hämäläinen and Olli Lagerspetz. I also want to thank the participants in the conference *Language, Ethics and Animal Life*, 26–28 March 2010, and the participants in the workshop *Questioning the Human Being*, Århus, 25–27 August 2011, for many valuable comments on earlier versions of this essay.

6

Talking about Emotions

Camilla Kronqvist

Are emotions internal states of the individual that can be identified in isolation from the particular cultural, social or interpersonal contexts in which they appear, or does the study of emotions require that we pay heed to these contexts if what we say is to be relevant for our understanding of emotions? In philosophy and the social sciences, different arguments for the latter position have dominated the debate in the past few decades. The new possibilities within neurosciences, however, have fuelled approaches to the study of emotions which attempt to explain emotions by referring to the biological, and often evolutionary, bases of emotion in our common human, or animal, nature.

One of the foremost proponents of biologically oriented views is Paul Griffiths. His much discussed work *What Emotions Really Are,* is a criticism of what he terms 'propositional attitude theories' about emotions. According to him, the form of conceptual analysis employed by these philosophers to gain clarity about emotions can only reveal 'a community's current beliefs' and stereotypes about what emotion words refer to.[1] These beliefs may well be mistaken, as people were mistaken in thinking that whales were a kind of fish.[2] The best answer to the question what emotions really are, Griffiths suggests, has to be informed by what current scientific research tells us about the referents of our emotion words.

Taking this route through the discussions of emotion research, Griffiths concludes that grouped under the term 'emotion' here are three distinct phenomena. First, there are *affect programmes*, including 'surprise, fear, anger, disgust (contempt?), sadness and joy.'[3] These are described as 'the coordinated set of changes that constitute the emotional response.'[4] They include: '(a) expressive facial changes, (b) musculoskeletal responses such as flinching

---

[1] 'Analysis can reveal the epistemic project in which a community is engaged with a concept, but the way that project develops will depend on what the community finds out through empirical investigation. If philosophers want to know about emotion, rather than about what is currently believed of emotion, analysis must proceed hand in hand with the relevant empirical sciences' (Paul E. Griffiths, *What Emotions Really Are* (Chicago, IL: Chicago University Press, 1997), 7).

[2] Ibid., 5, 7. See John Dupré, 'Natural Kinds and Biological Taxa,' *Philosophical Review,* 90 (1981), 759, for a discussion of whether science in this case really can be said to have shown that our former attitudes were based on a false categorization.

[3] Griffiths, *What Emotions Really Are*, 97.

[4] Ibid., 77.

and orienting, (c) expressive vocal changes, (d) endocrine system changes and consequent changes in the level of hormones, and (e) autonomic nervous system changes'.[5] Secondly, there are *higher cognitive emotions*. These constitute 'irruptive motivational complexes in higher cognition'[6] and lead us to depart from our long-term rational goals. Guilt, shame, vengefulness and (surprisingly) loyalty surface in the discussion of these. Thirdly, Griffiths delineates a group called *disclaimed actions* or *socially constructed emotions*, exemplified by phenomena such as being in love in the Western world or running amok in South-East Asia. On Griffiths' account, these are really pseudo-emotions which aim 'to take advantage of the special status that emotions are accorded because of their passivity . . . they are manifestations of the central purpose of higher cognitive activity – the understanding and manipulation of social relations'.[7] On his view, the 'affect programmes' constitute a clear category that should take the place of some of the phenomena we normally, but mistakenly for Griffiths, think of as emotions. He also argues that there is reason enough for introducing a second category that takes account of the higher cognitive emotions which do not coincide with affect programmes. He acknowledges, however, that more research is needed to work out what this category involves exactly.

The main targets of Griffiths' criticism of conceptual analysis are cognitive theories of emotion.[8] In this category he includes thinkers such as William Lyons, Anthony Kenny, Robert C. Solomon and Patricia Greenspan. My intention here is not to provide a defence of cognitive theories of emotion, nor provide a cognitive theory of my own. My reason for this is that cognitive theories are themselves faced with many problems. These include the tendency to intellectualize emotion at the expense of downplaying the importance of the spontaneous, embodied, character of emotion, as well as the attempt to systematize emotion, and language, to provide general rules for our use of emotion words.[9] However, I intend to show why I see it as essential to attend to the concept of emotion, and the concepts of different emotions in philosophical discussion. This in turn is a reason to consider the absolute necessity of a communicative context for being able to speak about and make sense of emotions. Here, I suggest it is of particular importance to consider what we *say* and *do* when we speak about emotions. First, however, there is reason to consider just how differently one in these approaches thinks of the contribution philosophy can make to the study of emotion.

1. There is at least one clear sense in which Griffiths misrepresents the aim of many cognitive *accounts* of emotions. By presenting the thinkers mentioned earlier as

---

[5] Ibid.
[6] Ibid., 245.
[7] Ibid.
[8] Ibid., 22–3.
[9] For Wittgensteinian criticisms of the limits of some forms of cognitivism, see: Phil Hutchinson, *Shame and Philosophy: An Investigation in the Philosophy of Emotions and Ethics* (Basingstoke: Palgrave Macmillan, 2008); Michael McEachrane, 'Capturing Emotional Thoughts: The Philosophy of Cognitive-Behavioral Therapy,' in *Emotions and Understanding: Wittgensteinian Perspectives*, ed. Ylva Gustafsson, Camilla Kronqvist and Michael McEachrane (Basingstoke: Palgrave, 2009) and Danièle Moyal-Sharrock, 'The Fiction of Paradox: Really Feeling for Anna Karenina,' in the same volume; and D. W. Hamlyn, 'The Phenomena of Love and Hate,' in his *Perception, Learning and the Self* (London: Routledge & Kegan Paul, 1983).

propositional attitude *theorists,* he treats their discussions as making theoretical, or hypothetical, claims that are not warranted by science. He takes them as proposing a research programme that has proved to be degenerative. If we follow his distinction between affect programmes and higher cognitive emotion, we could say that he equates the cognitive claim that emotions are beliefs or judgements, with the suggestion that all emotions function as the higher cognitive emotions. This, in his view, is unwarranted, since it denies the fact that there are affect programmes which precisely do not depend on any higher, or more complex, cognition. Jenefer Robinson makes a similar argument against cognitivism. She makes much of the fact that there are 'affective appraisals . . . that take place very fast and prior to any conscious cognition of complex information processing.'[10] She cites Joseph LeDoux who has 'discovered a fear circuit in the brain that operates very fast and without awareness.'[11] From the perspective of both Griffiths and Robinson, cognitive theories are seen as presenting scientific claims that have already been proven false.

Against this picture of propositional attitude theorists, one should remember that the philosophers presenting cognitive accounts of emotion take themselves to be speaking about the *logic, grammar* or *intelligibility* of emotion concepts – about, say, the relations between the concept of emotion and other concepts, such as those of belief, judgement and desire, value and character. They are concerned with providing an account of the *meaning* of emotion concepts where the meaning cannot be reduced to the referent of a word, or where the referent is only an aspect of a word's meaning. Although there is no clear agreement on exactly how one should think about investigating the meaning of a word, a central theme for many cognitive accounts is also to link questions about the meaning of emotions with questions that in a wide sense are moral.

Robert Solomon makes this explicit in an article addressing the different senses in which we may think of certain emotions as basic. He contrasts an Aristotelian notion of the way in which emotions are ' "basic" in terms of their significance in human life, and in particular in terms of ethics',[12] with the picture of searching for universal emotional building blocks, 'the crudest, least nuanced, least subtle and sophisticated emotions'.[13] The philosophers who attempt to define what emotions are with reference to something basic in the latter sense, thus, are guilty of stipulating that one use of 'basic' is more basic than others.[14]

In this respect, to analyse the current debates about emotion, it is not sufficient to discuss only what different answers different theories give to the question 'What

---

[10] Jenefer Robinson, 'Emotion: Biological Fact or Social Construction?' in Robert C. Solomon (ed.), *Thinking about Feeling: Contemporary Philosophers on Emotions* (Oxford: Oxford University Press, 2004), 35.

[11] Ibid.

[12] Robert C. Solomon, 'Back to Basics: On the Very Idea of "Basic Emotions"', *Journal for the Theory of Social Behaviour*, 32 (2002), 125.

[13] Ibid., 138.

[14] Since the writing of *What Emotions Really Are*, Paul Griffiths has also changed his view of what emotions are, in line with his outspoken aim to allow science to steer the way in finding a suitable definition. He recognizes the connections Solomon's writings have with socially situated, transactionalist views of emotions that, like Solomon, 'locate the primary significance of emotion in our interactions with other human beings' (Paul E. Griffiths, 'Emotion on Dover Beach: Feeling and Value in the Philosophy of Robert Solomon,' *Emotion Review*, 2 (2010), 27).

is an emotion?' It is as important to attend to the kind of question they consider themselves to be answering. What do their philosophical questions aim at and what is accepted as an answer? The real dispute in this debate is not about emotions, but about the meanings of words, and about how philosophy can contribute to our understanding of them. By suggesting that analysing our concepts of emotion only reveals our community's current beliefs about emotion, Griffiths endorses the view that there are no real philosophical questions beyond judging the power of scientific theories to explain the relation between empirical observations.[15] However, the force in distinguishing conceptual questions from empirical ones in emotion research, I suggest, with Peter Winch, lies in acknowledging that what people take themselves to be feeling is constitutive for identifying what it is they feel.[16] This leaves considerable room for the philosophical question of how we should understand the role concepts have in relation to what they purportedly describe.

In the following discussion I will show how the Wittgensteinian tradition opens up ways of understanding the question of the meaning of emotion words, as well as the nature of philosophy, that are often neglected in the current debate. In particular I want to approach this by discussing John Deigh's suggestion that cognitive theories of emotion fail to account for the ways in which, 'emotions are common to both humans and beasts'.[17] Among others, Deigh claims that the cognitivist's stress on intentionality, of forming beliefs as a basis for feeling emotions, makes it unintelligible to attribute emotions to animals. It presupposes the possession of linguistic capacities that would allow the animals to form beliefs, and clearly animals do not possess these capacities. Yet, our human emotions share an important ancestry with the other animals. The impetus of my discussion, therefore, is to advance an understanding of the intentionality of emotions, their 'aboutness' or 'directedness' towards a situation, which does not rely on the attribution of internal mental states to animals, where we immediately feel the need to find proof of such states.[18]

I also suggest that we should take a different attitude towards what it means to have a language. Rather than seeing language as the primary achievement of advanced mental

---

15   Stanley Cavell's essay 'Must We Mean What We Say?' is relevant to responding to these charges. Originally published in 1958, it is critical of the idea that the turn towards 'what we say' in ordinary language philosophy could be translated into what native speakers of English are inclined to say. See Stanley Cavell, 'Must We Mean What We Say?' in his *Must We Mean What We Say?* (Cambridge: Cambridge University Press, 2002).

16   Peter Winch, *The Idea of a Social Science and Its Relation to Philosophy* (London: Routledge and Kegan Paul, 1990 [1958]). For Winch this point extends beyond the use of emotion words to all phenomena that in one way or another can be called cultural or social. For a more recent contribution on the relevance of Winch's thinking, see Phil Hutchinson et al., *There Is No Such Thing as a Social Science: In Defence of Peter Winch* (Aldershot: Ashgate, 2008).

17   John Deigh, 'Primitive Emotions,' in *Thinking about Feeling: Contemporary Philosophers on Emotions*, ed. Robert C. Solomon (Oxford: Oxford University Press, 2004), 9.

18   Apart from Cavell and Winch, my discussion has great affinity with several discussions on the attributions of thought and emotion to animals in the Wittgensteinian tradition. These include: David Cockburn, 'Human Beings and Giant Squids,' *Philosophy*, 69 (1994); Norman Malcolm, 'Thoughtless Brutes,' in his *Thought and Knowledge* (London: Cornell University Press, 1977); Rush Rhees, *Moral Questions* (Basingstoke: Macmillan, 1999), 165–88; and Lynne Sharpe, 'Beyond the Pale,' in her *Creatures Like Us?* (Exeter: Imprint Academic, 2005).

processes, which provide the founding blocks for both reason and human culture, I will follow Pär Segerdahl's suggestion, after Wittgenstein, that it is more fruitful to see language as growing out of culture, as part of different ways of being and doing together.[19] The question whether animals, and in particular higher primates, can be said to have language, according to Segerdahl, has often focused on the formation of sentences and the production of intelligible sounds. It changes its form, however, if we consider the ways in which language is necessarily immersed in cultural forms of life, ways of acting and significantly *responding* to each other. To some extent we as humans share such responses with animals, especially those with whom we live in close contact and with whom we have come to share a form of life, say, within a household. The attribution of emotion to animals, in this respect, becomes an aspect of the sense in which we share a life. The question of what can be shared cannot be settled without attending to these contexts.

My following discussion proceeds in three steps. First, I show that the meaningful attribution of an emotion does not presuppose the existence of an inner state or process that the word refers to. Secondly, I consider whether the fact that we cannot extend all uses of emotion words to animals renders particular uses of the same word problematic. Thirdly, I offer reminders of the role the relationship we have to someone has for the meaning of the emotion we see in someone, as well as the relevance it has for using an emotional and mental vocabulary when speaking about an animal.

2. Suppose that I come home from work, and as I usually do, see to it that my cat gets food. When I open the door to the refrigerator, the cat comes running. Imagine, further – to provide a context in which I would see the need to comment on my cat's very normal reaction – that a friend is over for a visit, and is with me in the kitchen. I say to my friend: 'She always comes running when the refrigerator door opens,' or 'She knows she will get food when I come home,' or 'She's really hungry at this time of the day,' or 'She always wants to eat when I come home,' or 'She's always happy to get food.' How does this example help us become clear about the use of emotion words?

First, it reminds us of the connections between the concepts of knowing, thinking, wanting and feeling that cognitivists wish to bring out. In other words, it brings into view how having certain feelings is related both to having certain beliefs or making certain judgements and to having certain desires; for example, in the case of fear the judgement that 'it is dangerous' is related to the desire to run away. For this reason, I will persist in speaking in the same breath of what may easily be perceived as quite different mental states – such as beliefs, desires, emotions, physical sensations – although many cognitivists would probably find this problematic. Here, however, I concede to Griffiths the point that treating these concepts as delineating categories of isolable mental states where we are only asked to clarify their interconnections is attaching the wrong kind of seriousness to our everyday ways of speaking. It is treating these mental concepts

---

[19] See, for example, Pär Segerdahl et al., *Kanzi's Primal Language: The Cultural Initiation of Primates into Language* (Basingstoke: Palgrave Macmillan, 2005), and Pär Segerdahl, 'Humanizing Nonhumans: Ape Language Research as Critique of Metaphysics,' in this volume.

as expressive exactly of the kind of proto-scientific theories about the workings of the brain and body that Griffiths is correctly criticizing. By doing this one pays all too little attention to the possibilities of alternative ways of conceptualizing our emotional lives as well as to the different uses to which each of these concepts is subject in everyday speech.[20] Therefore, I am intent on showing how not making a clear distinction between these concepts may be informative for clarifying some concerns relating to emotions, and how at other times marking a distinction is worthwhile.

As the critics of cognitivism point out, it is certainly correct to say that my cat neither needs to be able to form the belief 'When the refrigerator door opens there will be food,' nor to do something similar to silently saying the words to itself, for me to legitimately attribute knowledge, happiness or hunger to it. What cognitive accounts do call attention to, however, are the internal relations between these descriptions of the situation; the ways in which they are mutually defining. By this I mean that, given a particular context, they *may* all serve as meaningful descriptions of my situation. I may say 'She knows she will get food . . .' or 'She's hungry . . .' or 'She's happy to get food . . .', and, in that context, it may make no difference which of these phrases I use. I may use the words interchangeably, or explicate what I mean by reference to one or several of these different descriptions. For this reason, one should be careful not to assume that these words necessarily apply to different inner states or processes.

The thought that 'knowing', 'being hungry' or 'happy' refers to different inner states, invites the temptation to think that these states can only be related causally. My speaking about the relation as an internal relation is meant to dissolve that idea. What allows us to speak of a connection between being hungry, wanting to eat or thinking that it would be good to have something to eat is not the perception of a causal connection between different processes. It is not that we have sufficient evidence of the fact that people who are hungry, perceived as a physical state, also have a desire to eat, perceived as a volitional state, or that the desire to eat gives rise to the thought, perceived as a cognitive process, 'It would be good to have something to eat.' Rather, in this case, being hungry *is* desiring to eat, and wanting to eat is having the thought that eating would be nice. This is what it *means* to be hungry and to want to eat; and it is internally related to the meaning of being happy or contented at having eaten.

This is not to say that it is unintelligible to say that I am hungry but do not want to eat. Suppose I have had the stomach flu. Then I may say that I am hungry in the sense that my stomach is contracting and I know it would be good for me to eat something, but that I still feel nauseated at the thought of food, and therefore do not want to eat. Similarly we may say of a baby that it is crying because it is hungry,

20 Cf. Anna Wierzbicka, 'Everyday Conceptions of Emotion: A Semantic Perspective,' in *Everyday Conceptions of Emotion: An Introduction to the Psychology, Anthropology and Linguistics of Emotion*, ed. James A. Russell et al. (Dordrecht: Kluwer, 1995). Wierzbicka remarks, for example (pp. 36–8), that universalist accounts of basic emotions, such as Paul Ekman's research that also inspires Griffiths' discussion of the affect programmes, run the risk of anglocentrism, by assuming that English words for emotions do indeed denote universal phenomena. Certainly *sadness* by contrast to *joy* seems to be the natural response to the loss of a loved one, but native speakers of other languages, and indeed English, may well feel more inclined to express their feelings in different terms in such cases.

judging by how long it has been since it last ate. Nevertheless, we may also say that it does not want to eat, since it turns away from its mother's breast. Thus, although the expressions I mentioned may be used interchangeably in some situations, they serve to make important contrasts in a different case. Just as I may say 'I know it's dangerous, but you needn't be afraid,' to reassure you of my confidence in being able to handle the danger, there is nothing in recognizing the internal relations between the different ways of expressing one's thoughts, desires and emotions which forces us to say any one thing in a particular case. What particular words I choose rather reflect the specific features of the situation to which I want to draw attention. If I want to comment on my cat's great appetite I say, 'She's always really hungry . . .' to point out how eagerly she strikes up against my legs while waiting to be served. If, by contrast, I want to comment on her intelligence, I say, 'She knows she'll get food . . .' Often, my choice of words also reveals a discrepancy in the different criteria to which I appeal in making a statement. The baby's crying is one criterion for hunger and wanting to eat. Its turning away from the breast, the one it would in many similar situations eagerly turn to, is a criterion for not wanting to eat. In this respect, my choice of words matters, but not because it allows me to pick out different inner states.

This need to distinguish between the different states a *person is in* may also suggest to us that some descriptions point to a real difference in what is being described. Again, however, I suggest we resist that temptation. The two points I want to take from this are the following. First, our thoughts and feelings are internally related in so far as we ask for an additional explanation in instances where they appear to be in contradiction, and do not ask for explanations when our thoughts are in line with our feelings and desires. If you are hungry we look for an additional cause or reason for not eating. Are you sick or on a diet? Your hunger, on the other hand, is your reason for wanting to eat, or being happy to have dinner. Secondly, the sense in which our particular choice of words matters to us in a situation is expressive of our different *concerns* in the situation, the contrasts we wish to mark out and so on. It does not depend on the essence of some underlying distinct phenomenon.

In this respect, despite criticizing the cognitivist for being too intellectualistic, Griffiths and Robinson still hold on to a too rigid and intellectualistic view of language. When Grifftiths, in his criticism of conceptual analysis, equates our use of concepts with a folk psychological theory, he suggests that our uses of psychological concepts are means of reporting inner states or explaining or predicting behaviour. Our emotion words are taken to denote states or processes, which themselves are taken to constitute underlying causes of behaviour. My point, in short, is that we fail to see the many things we *do* when we talk with each other, the different concerns our language of emotion gives expression to – also when we explain and predict behaviour – if we only think of language in this sense.[21]

---

[21] Cf. Ludwig Wittgenstein, *Philosophical Investigations*, rev. 4th edn (Malden, MA: Wiley-Blackwell, 2009), §§23–5.

The example also demonstrates the wrongness of supposing that some of these descriptions are necessarily more problematic than the others. One may think that in these meaningful descriptions of an animal we move from an easy case, describing outer behaviour, 'She comes running,' to a more difficult one ascribing physical needs, 'She is hungry,' to an even more difficult case of asserting 'She knows' or 'She is happy,' which involves the attribution of inner states that are necessarily hidden and whose existence, therefore, is more difficult to determine.[22] For this reason, some readers may object to my use of 'hunger' when discussing the attribution of emotions to animals, for who would deny that animals can feel hunger? To ease this concern, let us consider some of its possible sources.

First, the notion that some uses of our mental vocabulary are more problematic than others, as I present it here, relies on the suggestion that the attribution of knowledge, thoughts and happiness is dependent on inner states or processes for making sense. This, as I have already suggested, is not a presupposition we need to accept. Saying 'She's hungry' or 'She knows she'll get food . . .' does not by necessity have a different role in conversation than 'She comes running.' If the latter is conceived as unproblematic, the first may be as well. Saying this does not imply opting for a behaviourist conception of psychological concepts, according to which our descriptions of emotion and thought are reducible to descriptions of a creature's outer behaviour. I will not argue this point here, but if I did, I would point out how behaviour, too, in our normal description of it, is animated; it is seen in context as meaningful. Even in the case where we speak of a cat's behaviour we are talking about something other than the behaviourist's physiological changes. I am also not suggesting that speaking about the inner life of a person is meaningless.[23] What I criticize is rather taking such speech as referring to something that in fact is inner and thereby implying that there is an unbridgeable barrier to truly knowing what goes on with someone. On such a view, the inner lives of other people are by necessity hidden to us as a result of a supposed unbridgeable dualism between inner processes and outward behaviour. The philosophical work done in response to Wittgenstein's remarks on the impossibility of a private language, however, suggests that speaking of someone as having a rich inner life is itself dependent on things that in the right circumstances are possible for us to acknowledge.[24] A person is often silent, but harbours thoughts and feelings that find expression in a diary, or in intimate conversation. The realization that aspects of a person may be hidden to us, then, is connected with realizing that our lives together take a particular form. Openness about our state of mind is not always encouraged. People do not always say what they think. Our relations are not as close as we would wish. Trust often has

---

[22]   The notion of something hidden is of course compatible with trusting that science will reveal what is hidden in the future.

[23]   Iris Murdoch presents one of the most interesting readings of Wittgenstein's rejection of the idea of a private language in such terms. See her 'The Idea of Perfection,' in *Existentialists and Mystics*, ed. Peter Conradi (Harmondsworth: Penguin, 1999).

[24]   See, for example, Paul Johnston, *Wittgenstein: Rethinking the Inner* (London: Routledge, 1993), for an extensive discussion on what meaning speaking of the inner may have after Wittgenstein.

to be earned before confidences are shared, and so on.[25] It is against this background that the distinction between an inner and outer has sense.[26]

3. The latter remarks, however, help to spell out a different concern about the application of our emotional vocabulary to animals. Is it that the attribution of knowledge, thoughts or happiness is only metaphorical when we speak of a cat? Someone may want to suggest that it is not exactly meaningless or illegitimate to speak of knowledge and happiness in the cat's case, but that we do not, or *cannot*, use these two concepts in *a full sense* when the animal is concerned. The uneasiness one may feel at my use of hunger as an example of emotion relates to this concern. Whereas hunger is more akin to a physiological reaction, the attribution of an emotion such as happiness seems to demand a more complicated context, as well as an understanding of one's own situation's being of a certain kind, in order to make sense. Speaking of happiness in relation to animals, then, is stretching the word. 'Pleasure' or 'contentment' would be more fit expressions for the feeling of an animal after eating, for this form of satisfaction is far removed from the depths of human happiness. Such a suggestion appears to be the appropriate background for reading John Stuart Mill's well-known claim that '[i]t is better to be a human being dissatisfied than a pig satisfied; better to be Socrates dissatisfied than a fool satisfied'.[27]

Much speaks for taking such concerns seriously. Yet, if we are inclined to take this position, it is important to clarify to ourselves in precisely what sense we want to speak of 'using a word in a full sense'. Certainly, it is a significant aspect of our language of emotion that we may enter discussions about the *appropriateness* as well as the *depth* of an emotion. We enter discussions about what one should be happy about, we question whether happiness in a particular situation is the correct response or reveals a failure to respond compassionately to someone. *Schadenfreude* is here a case in point. Similarly, we criticize someone for not truly being happy, or for feigning happiness, even if someone has a reason to be happy. This is not only a matter of grading different forms of happiness or determining their appropriate objects; it is significant for these kinds of discourse that we frame our question in terms of *what* happiness is. That is, not

---

25   These descriptions of 'our' life are not meant to normatively delineate what human life necessarily should look like and to suggest that trust, for instance, is something that we necessarily have to earn. It is not that we cannot imagine a life where people in general were more trusting; after all, some people are. My point is only that a certain conception of an inner life is conceptually intertwined with other ways of speaking of, for example, trust or sharing confidence, or feeling as if one is somehow more closely connected to one person than to another.

26   The elaboration of this theme in Wittgenstein's thinking deserves a discussion of its own. İlham Dilman's lengthy discussions of these issues in several books, though often neglected within this context, strike me as insightful. In one place he remarks: 'The *philosophical* dichotomy of the inner and the outer in human life does not correspond to the variety of contrasts we make in our life using these or similar words.' Here 'the inner life' is at times 'a life in which a man turns reflectively to himself' to contrast with a life of action. At other times the contrast is used to distinguish a spiritual life from a worldly life, and yet at others to bring out a world of imagination – 'the world that is *in* his phantasies, the world *of* his phantasies, the world *as* he imagines it' (İlham Dilman, 'The Inner and Outer in Human Life,' in his *Love and Human Separateness* (Oxford: Blackwell, 1987), 60–1).

27   John Stuart Mill, *Utilitarianism* (London: Longmans, Green, 1882), 14.

only do we distinguish between good and bad forms of happiness, but we also suggest that it is good to be happy about certain things, or that happiness involves a certain way of looking at things. Furthermore, considerations such as these enter into our judgements about whether it is intelligible to speak of happiness in a particular case. Thus, our appreciation of what happiness is in itself introduces a certain understanding of what is good or valuable in human life, so that someone who does not embrace this understanding can be accused of being shallow or debased.

But how should we make sense of this seemingly necessary understanding of at least part of the concept of happiness? As we saw, cognitive theories are tempted to present the matter in such terms that a person needs to have the concept of happiness if we are to describe him or her as happy.[28] This tempts their critics to interpret emotions as hypothetical attributions of cognitive capabilities; something it could turn out that a person has or does not have. But, as we saw, determining whether someone can be said to be truly happy, or happy for the wrong reasons, does not primarily present us with a case of finding out whether certain cognitive processes are taking place. It contributes to a moral discourse of what is good and meaningful in life, what is profound or shallow, or of deep significance. This is the way that Solomon in reference to Aristotle spoke of certain emotions as being basic. It is also one way of understanding what it is for someone to suggest that we need to attend to these aspects of our life if we are to speak about happiness in a full sense. The notion of a full sense of the word is then expressive of a standpoint on life. However, the absolute character of this moral statement – 'This is what happiness is truly about!' – may also lead us to think that there is a full sense of using a concept that is philosophically misleading. This happens if we think that the word can be used to mean only one thing, or can be used in only one pure way, which renders all other uses disputable.

One way of dissolving this idea is by turning to Wittgenstein's remarks on family resemblance.[29] This, on my view, involves moving from an understanding of meaning as fundamentally referential, even if one word is taken to refer to different things, to an understanding of meaning as related to a range of interconnected uses of which some overlap but not necessarily all. On this view, there is no one singular feature that justifies us in speaking about 'knowledge' or 'happiness' in different conversations. Rather the words at different times draw attention to different aspects of a situation. On this note, the whole idea that some uses of words are more central, or more true to the meaning of a given word, is exactly that – an *idea*. Certainly, it is a powerful idea, and it easily and continuously shapes our philosophical thinking. In the end, however, it is not anything more than that, a background assumption that we may question and criticize once it is brought into light.

If we read Wittgenstein's remarks in this way, there are uses of 'knowing' or 'feeling happy' where what I say about a cat coincides with what I say of a grown up person or

---

[28]   See Richard Joyce, *The Evolution of Morality* (Cambridge, MA: MIT Press, 2007), esp. ch. 3, for a
       similar argument that does not accept the claims of cognitivism but still makes the possession of
       language central for moral emotion.
[29]   Wittgenstein, *Philosophical Investigations*, §§67–77.

a child. 'She comes running . . ,' 'She wants to eat . . ,' 'She is happy to get food,' 'Now she is content.' It is also worth noting how our inclination to see a cat as content after eating reflects our own feelings of contentment after having eaten in similar situations, and how in the attribution of different emotions to animals we take the animal to share our understanding of a situation. In so far as the situation offers us reasons for feeling certain ways – say, jumping at a loud bang, enjoying the warmth of the sun – we regard the animal as responsive to these same reasons. There are, however, uses of 'being happy' or 'knowing' which our life with a cat does not render meaningful. The sentence 'She knows I will be home later tomorrow, so she'll get her food herself,' becomes meaningful and true about my child at a certain age, and certainly is so of my wife. However, it has no bearing on my relationship with my cat. In other words, what I say about my cat is related to what it does, and can do, and to which reasons for feeling and acting I see the animal as being responsive.[30] (She comes running when the door opens, attacks the food when it is in the bowl and sniffs the package beforehand, but she does not open the refrigerator door, or the can containing the cat food, or pour the contents into the bowl herself.)

Or, again taking the example of happiness, the sense in which we say of a cat that it is happy to have had some food when it rolls up on the sofa and purrs contentedly afterwards, or is enjoying the sun, is, in many situations, quite the same sense in which we say that a human being is happy or contented after a good meal or when sun-bathing. However, this does not mean that we, to paraphrase Wittgenstein, can say of a cat that it feels happy at the prospect of eating fish tomorrow instead of meat, or that it is feigning happiness just to please us.[31] We do not accuse the cat of not being true to us. This, however, is an aspect of the life we share with other people and with animals; which of the things we say have or fail to have sense is expressive of those lives, and not of some cognitive or emotive abilities the cat has or does not have. From the point of view that I am here endorsing it is the life we share with an animal, our growing accustomed to its habits and character, which allows us to distinguish whether our cat is punching out its paw in anger or to play, or to spot our cat's intention, say, to jump up on the shelf where it usually takes a nap, whereas someone new to the cat may wonder why it is crouching on the floor.

If we take this view on language use, the suggestion that we do not speak of 'knowledge' or 'happiness' in a full sense when we speak about the cat running to the kitchen or feeling content after a meal, begins to waver. In these particular cases, this is what 'knowledge' and 'happiness' mean. Truly, we lose sight of many meaningful and important conversations if we assume that there are no other uses of 'knowing' and 'feeling happy' that can reveal significant aspects of these concepts. The fact that there are situations in which the words are also used in other ways, however, does not make

[30] In a similar manner, Rhees writes about the difference in speaking about taking responsibility for the course of one's life in relation to human beings and animals: ' "He was faced with a very difficult choice." That is something you would never say of an animal. Neither would you say that an animal showed weakness or strength in making a choice' (*Moral Questions*, 167).

[31] Cf. Wittgenstein, *Philosophical Investigations*, Part I, §250, and Part II, i, on whether a dog can be said to simulate pain or believe his master will come the day after tomorrow.

our use of the words in this narrow sense incomplete. There is no need to say that we do not use the words in a full sense. On the contrary, for this particular purpose our use of the words makes full sense.

The answers we are inclined to give to the question whether an animal can be said to fully know something or be happy, therefore, is expressive of our expectations concerning language. Someone who desires language to be orderly will react with discontentment at the often disorderly nature – or, more neutrally, open-ended character – of language, feeling it is a failure or a weakness. Someone, such as me, who sees the indeterminacy in meaning – the sense in which we can draw on different criteria to explain why we speak of knowledge or happiness in a particular case – as a strength in language, will not. When in philosophical conversation we are faced with questions such as 'Is this a case of an emotion or a non-evaluative belief?' there is a real possibility to, as Wittgenstein says, 'Say what you please, so long as it does not prevent you from seeing how things are. [With the important addition:] (And when you see that, there will be some things that you won't say.)'[32] This, of course, is not to say that scientists should give up striving to be systematic in their use of language within their research. For the particular purposes of their research questions, it may indeed be helpful to differentiate between different types of emotion as Griffiths does. This is important, not least in order to clearly define for oneself what is being investigated. What needs to be kept in mind, however, is that these scientific definitions always serve a distinct purpose, and that all forms of communication where a word is used do not necessarily share that same purpose. What we say in a concrete case is certainly not irrelevant. Its significance, however, is somewhere else than where we are first inclined to look; in *what* we do by choosing one word over another, and what we make of ourselves and others by doing so.

4. I already pointed out the significance that sharing a life with an animal has for the things we may say about that animal. In conclusion I want to consider a few more ways in which the *relationship* we have with someone is relevant for the meaning of our conversations about emotion. First, consider the fact that I often do not only speak about my cat, but also speak *to* it, although I do not expect a verbal response. I say, 'Oh, you're really hungry, aren't you?' and 'I have something good for you.' I say 'What have you done!' with anger at some mischief, or 'Ooh, that hurt!' when the cat presses its claws into me. By contrast with thinking that what I do here can primarily be put in the category of explaining and predicting behaviour by reference to an inner cause, my words can better be seen as natural extensions of my other ways of responding to my cat.[33] They are not a description of what I do, but part of my preparing food for the cat, cuddling it, jumping up at being clawed. Saying these things to my cat is an aspect of the relationship we share, sometimes part of a daily ritual, and expressive of my love for, and sometimes anger at, the cat and how it and its life matter to me. My

---

[32] Wittgenstein, *Philosophical Investigations*, §79.
[33] Cf. Wittgenstein's remark (*Philosophical Investigations*, §244) on how the child can be said to learn new pain behaviour when it learns to replace a cry of pain with an exclamation and, later on, with sentences such as 'It hurts.'

words embody a specific form of attention to my cat, and how attentive I am to it and what I pay attention to in its life is an aspect of me. Just as little as there is something hypothetical in my other ways of spontaneously responding to my cat, is there anything hypothetical in these ways of verbally responding to it.

Secondly, consider the different roles my speaking about my cat may have in a conversation. Speaking the way I did in my example in some situations constitutes just idle chatter. At times it is primarily a response to the situation, what the cat does, and a way of directing your attention to what we see, or confirming that you, in fact, see what I see. At other times it is a way of striking up a conversation. The significance of what the cat did in that sense gives way for other things we come to think of. We speak more generally of cats and pets, what there is for dinner, what happened at work, and so on. A conversation about my cat, however, may also be an invitation to an acquaintance to start talking about more personal things, as it says something about me and my life, and not just, say, about my work. My affection for my cat may even show a new side of me to someone who by then may realize what a limited view they had of me. I leave it open, what this new side brings out, since what I see, or project into, my cat may emphasize a trait that I find desirable and take pride in, such as intelligence, as well as features I find shameful or in other ways unattractive. I may identify with the characteristics of my cat, as well as situate myself in opposition to them. Yet, where one conversation about my cat is an invitation to address issues of a personal nature, another may be a way of avoiding a subject that I really need to address with a friend but which I feel is somehow too touchy or otherwise difficult to bring up. What is at stake here is not my affection for and connection with my cat, but my failure to connect and be in contact with my friend. Whatever truth we can assign to what I say about the cat, there is a different sense in which I fail to be true to my friend.

These reminders do not speak directly to the question of how we should understand animal emotional life, but they do speak of the very different kinds of conversation we take part in. These go well beyond fulfilling the function of explanation and prediction. Among them we find the conversations that so strictly follow social convention that we may accuse them of only presenting us with an empty form. We do, however, also find the conversations that touch us deeply, the intimate dialogues that bear testimony to the contact we experience in relation to one another, the times we speak from the heart or find it hard to speak in the first place.[34] It goes without saying that the significance and difficulties of these kinds of communication far exceed the demands on linguistic clarity in scientific research. These conversations themselves are expressive of emotion, serving to articulate what it is that we feel as well as what it is to which our feeling is a response. They remind us of the need to consider the significance that the relations in which we stand to each other have for the kinds of situation we face.

It is, as it were, not the case that my having a personal relation to you necessarily stands in the way of giving a fair description of what you feel, since what I say is coloured

---

[34] Dilman presents the failure to make such contact in our personal life as a form of affective solipsism. See Dilman, 'Affectve Solipsism and the Reality of Other People,' in his *Love and Human Separateness*.

by my own emotional responses, in part constituted by our mutual relationship. On the contrary, we usually think that being close to someone, and having experience of how another person has previously responded in a range of conversations, make us better equipped to answer the question of what the other person would say than if we had not previously been in contact with him or her. From this perspective, the suggestion that my own personal and emotional involvement with an animal leaves everything I say about its emotional life hopelessly anthropomorphic fails to make sense. No talk about emotion could get off the ground if it were not in any way related to our own emotional responses to another's (human or animal) emotions; feeling pain when someone is crying, joy when somebody else is laughing, and so on. This does not mean that all attribution of emotion is emotionally charged, but that attributions of emotion have meaning in the realm of these kinds of responses. Just as much is it a precondition for recognizing emotion in animals that we respond to animals in similar ways, and not only to animal emotion by analogy with human emotion, but to emotions in animals that do not take the same form in human life. Think, say, of recognizing joy in a horse rolling in the grass.[35] As much as we recognize human emotion in an animal, we also recognize in the animal our own animal nature. On this view, the one best equipped to answer the question of what a particular animal may feel, is not the one who has no personal relationship with the animal, but someone who is knowledgeable of animals in general, and is experienced and engaged in the interactions with this animal in particular.

Trying to answer the question of what an animal can feel in isolation from the concerns that inform a particular situation, and the *relevance* for us of a particular question such as 'Can an animal feel guilt?', may in that sense constitute the real deviation from the issues with which talking about emotion faces us. It is by seeing in what ways what is said in conversation matters to us that we see the relevant criteria for determining whether an utterance is true or not, not by entering a conversation with the conditions for settling the truth of any statement clear to us from the start. Rather than expecting that every question should be answerable by a 'true' or 'false', there is also reason to consider whether the question as it stands is meaningful, and if so what meaning it has.

We also stand in need of a much more sensitive grasp of the criticisms we can meaningfully direct at what is said about an animal's life. Some ways of presenting animals as human beings – in children's cartoons, for example – are so clearly anthropomorphic that they do not qualify for further reflection. They present animals as having concerns that are only meaningful in a particular form of human context. Yet, what of the descriptions of animals, within the context of their own life, that we consider sentimental, callous, misleading or inattentive? These kinds of criticisms draw on an emotional vocabulary that at the same time has a moral force. They do not aim at the fact that I am emotionally involved in the situation I share with an animal, for that involvement in itself is an aspect of the meaning my words have. Rather they question the character of my involvement and thereby also something of my own character. They challenge us to consider whether our words express genuine interest and involvement in the animal's life or, alternatively, merely a philosophically tedious emphasis on the impossibility of speaking about animals in certain ways.

---

[35]   See Cockburn, 'Human Beings and Giant Squids,' for more reminders in this direction.

7

Man as a Moral Animal: Moral Language-Games, Certainty and the Emotions

Julia Hermann

1. Introduction

In *On Certainty*, Wittgenstein states that he wants to conceive of the kind of certainty he is concerned with 'as something that lies beyond being justified or unjustified; as it were, as something animal'.[1] With the term 'animal' he points to the absence of reasoning: 'I want to regard man here as an animal; as a primitive being to which one grants instinct but not ratiocination.'[2] In this essay, I will explore to what extent there is 'something animal' in human morality.

Although the later Wittgenstein does not explicitly address ethical questions,[3] a number of interpreters ascribe a fundamental ethical concern to him.[4] Unlike them, I agree with Nigel Pleasants that there is 'no distinctively moral viewpoint . . . in Wittgenstein's later philosophy', but that this philosophy, and *On Certainty* in particular, 'can be of help in our thinking about ethics and ethical issues'.[5] In what follows, I will demonstrate a way of making this philosophy fruitful for questions of moral justification, moral competence, moral development and moral education.

In *On Certainty*, Wittgenstein famously claims that 'justification comes to an end'.[6] He is concerned with empirical, not moral, propositions, but I believe that his

---

[1] Ludwig Wittgenstein, *On Certainty*, ed. G. E. M. Anscombe and G. H. von Wright, trans. Denis Paul and G. E. M. Anscombe (New York: Harper & Row, 1972), §359.

[2] Ibid., §475.

[3] Wittgenstein did not address ethical issues in the writings that follow his 'A Lecture on Ethics' (*Ludwig Wittgenstein: Philosophical Occasions, 1912–1951*, ed. James C. Klagge and Alfred Nordmann (Indianapolis: Hackett, 1993), 115–55).

[4] For example Alice Crary, 'Wittgenstein and Ethics: A Discussion With Reference To *On Certainty*,' in *Readings of Wittgenstein's 'On Certainty',* ed. Danièle Moyal-Sharrock and William H. Brenner (Basingstoke: Palgrave Macmillan, 2005), 275–301; Cora Diamond, 'Ethics, Imagination and the Method of Wittgenstein's *Tractatus*,' in *The New Wittgenstein*, ed. Alice Crary and Rupert Read (London: Routledge, 2000), 149–73; Stephen Mulhall, 'Ethics in the Light of Wittgenstein,' *Philosophical Papers*, 31 (2002), 293–321.

[5] Nigel Pleasants, 'Wittgenstein, Ethics and Basic Moral Certainty,' *Inquiry*, 51 (2008), 242.

[6] Wittgenstein, *On Certainty*, §192. *On Certainty* consists of notes which Wittgenstein wrote in the last two years of his life. He did not have time to revise them. Someone who tries to interpret these unrevised notes is skating on thin ice. Nonetheless, I have no doubt that they contain too many insights to simply disregard them.

reflections on how justifications come to an end with regard to empirical propositions can illuminate the status of moral propositions such as 'killing is, *ceteris paribus*, wrong'.[7] Wittgenstein's insight that doubt does not always make sense has a bearing on the question as to whether all moral 'beliefs' can be subject to doubt.[8] That we cannot carry on justifying indefinitely is true of moral justification as much as it is true of empirical justification.

I will start by arguing for an understanding of morality as a conglomerate of overlapping language-games. I will then look at Wittgenstein's description of what could be called 'empirical certainty', and draw an analogy between this kind of certainty and what we could call 'moral certainty'.[9] It will be argued that moral certainty can to a certain extent be regarded as 'animal', and that it is closely related to what I will call 'moral competence': it is for morally competent persons that the moral wrongness of certain types of action and the rightness of others is certain. I will explain what I mean by 'moral competence', drawing particular attention to the emotional capacities it involves in Section 5. In Sections 6 and 7, I will point out the effect animals have on moral development. Finally, I will point out the animal as well as the human dimensions of our morality.

Although I use the rather technical term 'moral competence', it is important to note that I do not think of moral persons as possessing any special knowledge of facts, or as having mastered any particular technique. In the absence of a better word, I talk about competence in order to highlight that moral agency requires numerous capacities that have to be developed through experience and practice.

2. Moral language-games

When Wittgenstein describes different language-games he does not include activities of moral evaluation, morally praising and blaming, holding someone to account, etc. Rather, he focuses on things like '[g]iving orders, and obeying them', '[d]escribing the appearance of an object, or giving its measurements', '[r]eporting an event' and '[m]aking a joke'.[10] However, there is no good reason for restricting the metaphor of the

---

[7] I use 'ceteris paribus' in the sense of 'absent special circumstances'.

[8] I talk about moral 'beliefs' here, although the 'beliefs' concerned differ from beliefs like the belief that it is raining. That killing is, ceteris paribus, wrong, is arguably rather a kind of attitude than a belief. 'Belief' here should therefore not be understood in a strict sense.

[9] I am aware of two other attempts to draw such an analogy. These are the attempts by Michael Kober and Nigel Pleasants. See Michael Kober, 'On Epistemic and Moral Certainty: A Wittgensteinian Approach,' *International Journal of Philosophical Studies*, 5 (1997), 365–81; Pleasants, 'Wittgenstein, Ethics and Basic Moral Certainty'; and Pleasants, 'Wittgenstein and Basic Moral Certainty,' *Philosophia*, 37 (2009), 669–79. These two versions of the analogy differ from each other, and both of them differ from the one presented in this essay. I discuss Pleasants' and Kober's analogy in Julia Hermann, *Being Moral: Moral Competence and the Limits of Reasonable Doubt*, PhD thesis (Florence: European University Institute, 2011), ch. 4.

[10] Ludwig Wittgenstein, *Philosophical Investigations*, trans. G. E. M. Anscombe, 3rd edn (Oxford: Blackwell, 1968), §23.

language-game to the non-moral uses of language. If it is illuminating at all, it has to be applicable to the moral dimension of human life and language.

I will thus conceive of our moral uses of language and the actions with which they are intertwined in terms of (moral) language-games.[11] Examples of such language-games, to which I will also refer as 'moral practices', include various types of activities and behaviour, both verbal and non-verbal, such as: 'holding someone responsible', 'providing a justification for the breach of a promise', 'discussing whether or not we have moral obligations towards animals', 'consoling someone in distress', 'helping someone', 'deliberating about the right action in a given situation'. We cannot single out one feature by virtue of which these activities are all 'moral language-games', but they reveal family resemblances.[12] Unlike Pleasants, whose understanding of *On Certainty* and its implications for ethics is otherwise close to mine, I don't restrict such games to 'belief, judgement, and reflection'.[13] Moral uses of language are associated with non-verbal reactions and attitudes, for example with reactions that have an emotional core such as shame or guilt.[14]

An understanding of morality as a conglomerate of overlapping language-games contrasts with a conception of it as a set of principles from which judgements about particular actions or situations can be derived by deduction. On the former account, morality is a practice and as such constituted not only by certain rules, but also by its 'surroundings',[15] its functions and its 'point'.[16] That the constituents of practices or language-games involve more than rules is convincingly argued by Timo-Peter Ertz. Ertz explains what it means that the 'surroundings' of a practice have a constitutive role by reference to the practice of weighing.[17] It is part of the surroundings of that practice that pieces of cheese do not suddenly grow. Moral language-games are partly constituted by the vulnerability of human beings, their dependence on others, the fact that there is a scarcity of goods and so forth. Just as the 'procedure of putting a lump of cheese on a balance and fixing the price by the turn of the scale would lose its point if it frequently happened for such lumps to suddenly grow or shrink for no obvious reason',[18] jumping into a lake fully dressed and taking a child out of it would lose its point if the water could not harm the child. The procedure Wittgenstein describes

---

[11] Compare ibid., §7.

[12] That they reveal family resemblances means that they exhibit 'a complicated network of similarities overlapping and criss-crossing' (ibid., §66).

[13] Pleasants, 'Wittgenstein, Ethics and Basic Moral Certainty', 263.

[14] Due to limits of space I will not address a number of questions connected to my reference to 'our moral language-games', such as the question as to who can be said to play these games, the extent to which the moral language-games played far away from us or at some time in the past differ from those we play today, whether and how we can morally criticize practices common in other cultures, etc. I deal with these questions in Hermann, *Being Moral: Moral Competence and the Limits of Reasonable Doubt*, ch. 7.

[15] Timo-Peter Ertz, *Regel und Witz: Wittgensteinsche Perspektiven auf Mathematik, Sprache und Moral* (Berlin: de Gruyter, 2008), 37, my trans. The German term he uses is 'Umgebung'. Wittgenstein uses 'Umgebung' in *On Certainty*, §350, where he refers to the circumstances in which a sentence is used.

[16] Ibid., 10f. 'Point' is my translation of Ertz' notion 'Witz'.

[17] Ibid., 37.

[18] Wittgenstein, *Philosophical Investigations*, §142.

would under those circumstances not be an act of weighing, and the act of jumping into a lake fully dressed and taking a child out of it would not be an act of rescuing a child.

However, the facts that are constitutive of moral language-games as part of their surroundings do not justify these games. Nor do the functions of morality play a justificatory role. In terms of justification, practices are autonomous, as Ertz argues.[19] It follows that on the account of morality presented here, attempts to base morality on non-moral premises are deeply flawed.[20] Human nature, to which some philosophers want to reduce morality, plays a constitutive role, but not a justificatory role. This is an important difference. Facts about our human nature are not the *reason* for killing being, ceteris paribus, wrong. We cannot justify our firm conviction that killing is wrong by referring to any of these facts. Nevertheless, there is a fundamental connection between those facts and our moral attitudes.

3. Empirical certainty

In *On Certainty*, Wittgenstein reflects on cases in which something is certain for us in the sense that we do not know what a doubt concerning this state of affairs would look like, and are at the same time unable to provide reasons for its truth that are more certain than what they are supposed to justify.[21] The various examples he gives include the existence of one's hands, the existence of the earth prior to one's birth, what one's name is and the identity of a familiar object (e.g. a tree) perceived under favourable conditions.[22] He establishes that the states of affairs concerned (that my name is J. H., that *this* is a tree, etc.) are not objects of knowledge, but of certainty.[23] In order to be knowable, they would have to be open to doubt and justification, which they are not. Doubt has to be based on reasons, and in the cases Wittgenstein looks at reasons for doubt are not available.[24] Moreover, in order for them to be possible objects of knowledge, it would have to be possible to be mistaken about them.[25] Yet it is impossible for me to be mistaken about the fact that the earth existed long before I was born, or about my name.[26] If in such cases it turned out that what I believed was false, this could not be fitted into the things I know.[27] 'Being mistaken' is not the same as 'saying something false'. We can imagine exceptional cases in which someone is in

[19] Ertz, *Regel und Witz: Wittgensteinsche Perspektiven auf Mathematik, Sprache und Moral*, 9.
[20] Examples of such attempts include David Gauthier, *Morals by Agreement* (Oxford: Clarendon Press, 1986); Peter Stemmer, *Handeln zugunsten anderer: Eine moralphilosophische Untersuchung* (Berlin: de Gruyter, 2000); Michael von Grundherr, *Moral aus Interesse: Metaethik der Vertragstheorie* (Berlin: de Gruyter, 2007).
[21] See Wittgenstein, *On Certainty*, §§4, 111, 250, 391, 445 and 470.
[22] See ibid., §§40, 84, 576 and 585.
[23] See, for example, ibid., §§151 and 194.
[24] See ibid., §§4, 122 and 458.
[25] See ibid., §674.
[26] See ibid., §155.
[27] See ibid., §74.

fact mistaken about his name, but it holds for the vast majority of people that such a mistake is impossible.[28]

Wittgenstein's reflections are prompted by G. E. Moore's attempt to prove the existence of the external world via his alleged knowledge of the existence of his hands, and by Moore's claims to know certain things with certainty.[29] Wittgenstein criticizes Moore's use of 'I know', arguing that what he claims to know, for example that *this* is his hand, is certain for him in the sense described above. According to Wittgenstein, the sentences which express these 'certainties' are true only inasmuch as they are 'an unmoving foundation of [our] language-games'.[30] They seem to express empirical propositions, but this appearance is misleading.[31]

While Wittgenstein thus seems to argue that our justifications stop at the point where we reach propositions which are not open to reasonable doubt, he also makes the apparently incompatible claim that the end of justification is *not* a proposition: 'Giving grounds, however, justifying the evidence, comes to an end; – but the end is not certain propositions' striking us immediately as true, i.e. it is not a kind of *seeing* on our part; it is our *acting*, which lies at the bottom of the language-game.'[32] Or as he formulates it in another paragraph: 'As if giving grounds did not come to an end sometime. But the end is not an ungrounded presupposition: it is an ungrounded way of acting.'[33] This is echoed by variations of the phrase 'This is simply how we act.'[34]

There thus appears to be a tension between Wittgenstein's remarks about certain propositions and his remarks about the fundamental acting. Since Wittgenstein did not have time to revise the notes published as *On Certainty*, it has been speculated that *On Certainty* might contain two inconsistent views, or that it might reflect a development in Wittgenstein's thought from a view focusing on propositions towards a view focusing on action.[35] My impression is that there is no real tension between the remarks about propositions that cannot be doubted and the passages in which Wittgenstein emphasizes that the language-games we play are grounded in action. The apparent tension disappears if we understand the underlying acting as both linguistic and pre-linguistic.[36] As a competent speaker of the English language and a competent participant in epistemic practices, I would not doubt, under normal circumstances, that my hands exist. I use my hands and refer to them verbally without any hesitation, but at the same time without justification. What lies at the root of the epistemic

---

[28] I thank the editors for stressing this point.

[29] G. E. Moore, 'Proof of an External World' and 'A Defence of Common Sense,' in his *Philosophical Papers* (London: Allen and Unwin, 1959).

[30] Wittgenstein, *On Certainty*, §403.

[31] See ibid., §308.

[32] Ibid., §204.

[33] Ibid., §110.

[34] See, for example, Wittgenstein, *Philosophical Investigations*, §§211 and 217.

[35] The first position has been defended by Michael Williams, the second by Avrum Stroll. See Michael Williams, 'Wittgenstein, Truth and Certainty,' in *Wittgenstein's Lasting Significance*, ed. Max Kölbel and Bernhard Weiss (London: Routledge, 2004), 247–81; Avrum Stroll, *Moore and Wittgenstein on Certainty* (New York: Oxford University Press, 1994).

[36] I argue for this view in Hermann, *Being Moral: Moral Competence and the Limits of Reasonable Doubt*, ch. 2.

language-games I participate in is my exercise of epistemic and linguistic competence. On this view, the proposition '*This* is a hand' formulates a fixed point of these practices. It expresses something that is fixed or certain for me *qua* competent epistemic agent and competent speaker of the English language. This way of solving the tension differs from the view that the certainties concerned are full-blown 'animal certainties' and as such purely pre-linguistic, like instincts.[37] At the same time, it also differs from the view that the underlying acting should be identified with linguistic competence.[38]

4. Moral certainty

Let us now turn to our moral language-games. Here, the wrongness of killing as such – that is, in the absence of special, justifying circumstances such as circumstances in which killing someone is the only way of defending oneself – is not disputed.[39] It seems that the sentence 'Ceteris paribus, killing is wrong' cannot be meaningfully questioned and justified on moral grounds, within our moral language-games. While we have disputes, for instance, about the legitimacy of abortion or euthanasia, the wrongness of killing as such is 'a hinge on which [our] dispute can turn'.[40]

Propositions such as 'Abortion is morally prohibited' and 'Abortion is morally permissible', by contrast, do not have this status. People come up with arguments for and against practices such as abortion and euthanasia, but we do not know what reasons to give to someone who denies that there is anything wrong with killing as such. Such a person is obviously not a competent player of our moral language-games.[41] Mastery of these games is incompatible with a serious denial of the moral wrongness of harming or killing people in the absence of justifying circumstances.

We can thus say that just as under normal circumstances a competent *epistemic* agent cannot reasonably doubt the existence of his hands, a competent *moral* agent has, in the absence of justifying excuses such as the need to defend herself, no reason to doubt the wrongness of killing. While the existence of our hands is certain for us insofar as we are *epistemically* competent, the wrongness of killing is certain for us insofar as we are *morally* competent. Accordingly, 'Ceteris paribus, killing is wrong' is true only inasmuch as it is an unmoving foundation of our moral language-games.[42]

---

[37]   Cf. Danièle Moyal-Sharrock, *Understanding Wittgenstein's 'On Certainty'* (Basingstoke: Palgrave Macmillan, 2004), 8.

[38]   This is the interpretation of Patricia Hanna and Bernard Harrison. See Patricia Hanna and Bernard Harrison, *Word and World: Practice and the Foundations of Language* (Cambridge: Cambridge University Press, 2004), 188.

[39]   See Thomas M. Scanlon, *What We Owe to Each Other* (Cambridge, MA: Harvard University Press, 1998), 200.

[40]   Wittgenstein, *On Certainty*, §655.

[41]   Here again the technical sounding terminology should not convey the picture of the moral person as one who has mastered the technique of playing a particular, well-defined game. Rather, I suggest thinking of her as being able to participate in many overlapping practices, which lack sharp boundaries.

[42]   Compare Wittgenstein, *On Certainty*, §403.

I suggest conceiving of the propositions concerned as the formulations of fixed points of our moral practices, which are abstracted from these practices ex post. The ways in which I talk and act reveal not even the slightest doubt regarding my name, the existence of my hands or that of the earth before I was born. Analogously, my behaviour towards others, my moral judgements, my expectations from others and myself, my moral justifications and what I accept as a moral reason reveal no doubt regarding the ceteris paribus wrongness of killing or abusing people, the ceteris paribus rightness of helping or rescuing or the ceteris paribus goodness of love and the ceteris paribus badness of hate.

But there seems to be a disanalogy here: While no sane adult acts in ways which reveal doubts regarding the existence of his hands etc., many actions of apparently sane people seem to be interpretable as revealing doubts and even a rejection of, for example, the prohibition of killing.[43] However, in cases where a person's behaviour over time reveals such a rejection, we usually have doubts regarding her sanity, for example in cases of criminals who continuously violate moral rules without exhibiting any feelings of guilt. Such people are called 'psychopaths' or 'sociopaths', and many of us think that they are in need of psychiatric care rather than imprisonment. Moreover, not every action that is regarded as immoral or morally prohibited should make us suspicious of the moral competence of the actor performing the action.[44] More than one single action needs to be considered in order to judge whether or not an actor is (fully) morally competent. The feelings associated with immoral actions are particularly telling in this regard. It also has to be taken into account that moral competence comes in degrees.

I suggest understanding the unjustified ways of acting at the root of our *moral* language-games as involving both linguistic and pre-linguistic action: ways of using moral concepts naturally and unreflectively, and ways of unhesitatingly helping other people, taking their interests into consideration and caring about them. These ways of acting express an 'animal certainty' in the sense that they do not involve a process of reasoning. That human beings are capable of reasoning and do have reasons for many things they do and think does not imply that everything is done on the basis of a reason or can be given a justification. Reasons play a prominent role in moral language-games, but in playing these games, we do, say and feel many things without a justification, yet not wrongfully.[45]

More about the non-rational components of morality will be said below. I take it that our moral certainties can also be said to be 'animal' in the sense that we take up the moral standpoint without justification. From the moral point of view, we are

43   I thank a participant in the First Nordic Wittgenstein Society Conference, at which a first version of this essay was presented, for drawing my attention to this point.
44   Drawing the line between cases of moral incompetence and cases of evilness is of course extremely difficult. It is even debatable whether there are any cases of serious wrongdoing which are not reducible to cases of incompetence.
45   Compare Wittgenstein, *Remarks on the Foundations of Mathematics*, part VII, §40 and Wittgenstein, *Philosophical Investigations*, §289. There Wittgenstein says that using a word without a justification does not imply using it wrongfully.

able to provide reasons for many moral judgements we make, but that as morally competent agents we take up this point of view in the first place is not the result of a process of reasoning. Just like language, morality 'did not emerge from some kind of ratiocination'.[46]

5. Moral competence and its emotional components

A fully morally competent person has mastered the many different, overlapping moral language-games we play. As a competent player of these games, she has a number of capacities, including a capacity for moral judgement, autonomy, empathy and feeling guilt when this is appropriate. She is able to take the interests of others into account for their own sake, to put herself in their shoes, to imagine the consequences her actions can have for others, and to feel what it is like for another to be harmed. In short: she is able to take a moral point of view.

Taking this point of view means not only thinking, but also *feeling* in moral terms. It means to feel pity for someone who suffers, to in some sense feel the other person's pain, to feel guilty for having broken a promise in the absence of excusing circumstances, to feel shame for having disappointed a good friend, and to feel indignation in the face of a harm done by someone else to a third person. It also means to be motivated to help, to reduce someone else's suffering, to keep a promise, to tell the truth and so forth.

The capacities mentioned above are not all distinct faculties. The capacity for moral judgement for example involves a capacity for empathy. Recent experiments conducted by cognitive neuroscientists and psychologists show that reason and feeling work together in moral judgement.[47] Many of our moral judgements appear to be correlated with automatic emotional responses, involving ancient areas of our brains, not foremost the prefrontal cortex. Research on psychopathology further supports the view that emotional capacities are crucial for the exercise of moral competence.[48] Drawing particular attention to the emotional components of moral competence, which tend to be overlooked by philosophers from the rationalist camp,

---

[46] Wittgenstein, *On Certainty*, §475. The German original reads: 'Die Sprache ist nicht aus einem Raisonnement hervorgegangen.' I present the analogy in more detail in Hermann, *Being Moral: Moral Competence and the Limits of Reasonable Doubt*, ch. 4.

[47] See, for example: Liane Young et al., 'Damage to Ventromedial Prefrontal Cortex Impairs Judgment of Harmful Intent,' *Neuron*, 65 (2010), 845–51; Liane Young et al., 'Disruption of the Right Temporoparietal Junction with Transcranial Magnetic Stimulation Reduces the Role of Beliefs in Moral Judgments,' *PNAS*, 107 (2010), 6753–8; Joshua D. Greene et al., 'An fMRI Investigation of Emotional Engagement in Moral Judgment,' *Science*, 293 (2001), 2105–8; Joshua D. Greene and Jonathan Haidt, 'How (and Where) Does Moral Judgment Work?', *Trends in Cognitive Sciences*, 6 (2002), 517–23; Jonathan Haidt, 'The Emotional Dog and its Rational Tail: A Social Intuitionist Approach to Moral Judgment,' *Psychological Review*, 108 (2001), 814–34.

[48] R. James and R. Blair, 'A Cognitive Developmental Approach to Morality: Investigating the Psychopath,' *Cognition*, 57 (1995), 1–29; R. James and R. Blair, 'Moral Judgment and Psychopathy,' *Emotion Review*, 3 (2011), 296–8.

my account is in line with the Aristotelian tradition, which stresses the importance of 'affective dispositions' and the role of 'sentimental education'.[49] The empirical research I am referring to can, I believe, contribute to a better understanding of our moral practices. It should not, however, be taken to fully explain or to justify these practices.[50]

6. Moral learning

The ability to take the moral point of view cannot be the result of the prescriptive teaching of moral principles, which is at the centre of rationalist accounts of moral education. Human beings develop moral competence in practice, through interactions with others and notably through the experience of conflicts. The child learns to treat others with respect, to take their interests into account for their own sake, to put herself in their shoes, to make moral judgements and to react with feelings of guilt, shame, indignation or pity in situations where this is appropriate. When a particular moral feeling is regarded as appropriate varies (to a certain extent) between societies. Although it seems to be part of the natural endowment of human beings that they have at least a rudimentary disposition to respond with such feelings, the development of the respective affective dispositions requires education of the sentiments as part of moral education, as well as (parental) love and care.

Moral education is undertaken not only by one's parents, but also by teachers, friends and ultimately by society as a whole. To a large extent, it is not carried out by any particular person. Rather, social interactions provide a training ground on the basis of which moral competence is developed. The numerous different capacities involved in moral competence are developed by way of experiencing concrete situations and interacting with others, not merely with human beings but also with animals. The sociologist Michael Rustin emphasizes that the 'essence of moral learning is not intellectual subscription to abstract precepts, but a process of learning-within-a-situation, from experience and example, in which the implications and effects of feelings and actions can be reflected on with others.'[51]

If we ask ourselves how the relevant emotional capacities might be developed, it seems plausible to conceive of moral learning as often starting with concrete situations in which our acting meets resistance. It seems that the experience of conflict and of being blamed for one's behaviour, as well as the experience of the emotional reactions

---

[49]   Jan Steutel and Ben Spiecker, 'Cultivating Sentimental Dispositions Through Aristotelian Habituation,' *Journal of Philosophy of Education*, 38 (2004), 532.

[50]   I am grateful to the editors for encouraging me to make this point explicit.

[51]   Michael Rustin, 'Innate Morality: A Psychoanalytic Approach to Moral Education,' in *Teaching Right and Wrong: Moral Education in the Balance*, ed. Richard Smith and Paul Standish (Staffordshire: Trentham, 1997), 87.

of others to one's own actions is important for developing certain moral feelings; that is, for feeling guilt, shame, remorse, etc. in the appropriate situations.[52]

A form of moral education suitable for the development of emotional capacities is what Martin Hoffman refers to as 'induction'.[53] Induction involves for instance reacting to cases in which a child hurts another child by making the child imagine what it would feel like to experience similar harm. It 'highlights both the victim's distress and the child's action that caused it and has been found to contribute to the development of guilt and moral internalization in children'.[54] The moral sentimentalist Michael Slote adds to Hoffman's account of induction that parents using this mode of moral education thereby 'demonstrate an empathic concern for the child (let us assume) who has been hurt by their own child, and there is in fact no reason why a child can't take in such an attitude, such motivation, directly from a parent'.[55] Inductive training thus involves also a kind of modelling: 'in most cases, induction will involve not only a parent's deliberately making a child more empathically sensitive to the welfare or feelings of others but also the child's directly taking in, by a kind of empathic osmosis, the parent's own empathic concern for others'.[56] This 'empathic osmosis' can also be expected to occur at times when induction is not used and a child simply notices a parent's empathic concern and attitude of care. We are dealing here with a kind of modelling which differs from that advocated by most educational theorists by being non-deliberate and possibly unconscious. Both induction and modelling are important ways of strengthening empathic concern for others.[57]

These theoretical arguments concerning moral education (as well as those presented in the following section) should not be understood as grounding the moral practices we come to participate in. According to the view presented in this essay, moral practices are not grounded in any form of argument or reasoning. Nevertheless, we can reason about how these practices work and how we learn to participate in them, and make arguments for particular ways of assisting children in becoming moral.

7. The role of animals in moral education

As a body of relatively recent empirical research shows, another way of strengthening empathic concern is to enable children to interact with – and take care of – animals.

[52] I discuss the role sanctions have for moral development in Julia Hermann, 'Die Praxis als Quelle moralischer Normativität,' in *Moral und Sanktion: Eine Kontroverse über die Autorität moralischer Normen*, ed. Eva Buddeberg and Achim Vesper, third volume of the series *Die Herausbildung normativer Ordnungen*, ed. Rainer Forst and Klaus Günther (Frankfurt and New York: Campus, forthcoming).

[53] Martin L. Hoffman, *Empathy and Moral Development: Implications for Caring and Justice* (Cambridge: Cambridge University Press, 2000), 10.

[54] Ibid.

[55] Michael Slote, *Moral Sentimentalism* (Oxford: Oxford University Press, 2010), 20. Moral sentimentalism is a meta-ethical view according to which morality is somehow grounded in emotions ('moral sentiments' in the terminology of philosophers such as David Hume and Adam Smith). It is an alternative to rationalism, which is the dominant position.

[56] Ibid. Slote draws on Hume's belief that not only basic feelings, but also opinions and attitudes can 'spread to others by contagion or infusion'. He refers to: David Hume, *Treatise of Human Nature*, ed. Selby-Bigge (New York: Oxford University Press, 1978), 320–4, 346, 499, 589, 592 and 605.

[57] Slote, *Moral Sentimentalism*, 20–1.

There exists empirical evidence for the positive effects animals have on moral development, and in particular on a child's development of empathy.[58] Therefore, it would be desirable that educators do more to encourage and enable children to take care of animals.[59]

It has been found that having a pet as a child correlates with greater concern towards humans later on in life.[60] Given that not every child has the possibility of keeping a pet at home, primary school teachers could provide the possibility for interaction with animals, for example by having pets in the classroom.[61] Even though research in this area is still fairly limited, it nevertheless 'consistently indicates the positive benefits of having animals in educational environments, particularly dogs'.[62] Not only live animals, but also discussions about and representations of animals seem to foster the moral development of children.[63] Pupils do not merely learn to treat animals well, which is typically the goal of 'humane education' programmes,[64] but also come to be more empathic towards each other.[65] Primary school teachers who have animals in the classroom and encourage their pupils to interact compassionately with them report that the animals have a positive effect on the pupils' social behaviour, including a decrease in aggressive behaviour.[66]

In their study on teachers' experiences with humane education and animals in the elementary classroom, Beth Daly and Suzanne Suggs quote the following reflection of a teacher: '[An animal] teaches the children . . . compassion, responsibility, caring, love and brings out the innate abilities in some to show feelings that most sometimes hide behind to protect themselves.'[67] On the basis of their own study and those of others, the authors conclude that 'there appear to be a wealth of emotional benefits reaped by students from the simple endeavour of keeping a pet in the classroom'.[68] They believe

---

[58] See Alan Beck and Aaron Katcher, *Between Pets and People: The Importance of Animal Companionship* (West Lafayette, IN: Purdue University Press, 1996); June McNicholas and Glyn M. Collis, 'Children's Representations of Pets in Their Social Networks,' *Child: Care, Health, and Development*, 27 (2001), 279–94; Gail F. Melson, *Why the Wild Things Are: Animals in the Lives of Children* (Cambridge, MA: Harvard University Press, 2001).

[59] See Beth Daly and Suzanne Suggs, 'Teachers' Experiences with Humane Education and Animals in the Elementary Classroom: Implications for Empathy Development,' *Journal of Moral Education*, 39 (2010), 101–12.

[60] See Elizabeth S. Paul and James A. Serpell, 'Childhood Pet Keeping and Humane Attitudes in Young Adulthood,' *Animal Welfare*, 2 (1993), 321–37.

[61] It might, however, often be impossible to have a pet in the classroom, since many children suffer from allergies. I thank the editors for making me aware of this problem.

[62] Daly and Suggs, 'Teachers' Experiences with Humane Education and Animals in the Elementary Classroom: Implications for Empathy Development,' 110.

[63] Ibid., 102.

[64] See ibid.

[65] See Frank R. Ascione and Claudia V. Weber, 'Children's Attitudes about the Humane Treatment of Animals and Empathy: One-Year Follow Up of a School-Based Intervention,' *Anthrozoos*, 9 (1996), 188–95.

[66] See Andreas Hergovich et al., 'The Effects of the Presence of a Dog in the Classroom,' *Anthrozoos*, 15 (2002), 37–50; Kurt Kotrschal and Brita Ortbauer, 'Behavioral Effects of the Presence of a Dog in a Classroom,' *Anthrozoos*, 16 (2003), 147–59.

[67] Daly and Suggs, 'Teachers' Experiences with Humane Education and Animals in the Elementary Classroom: Implications for Empathy Development,' 108, my ellipsis.

[68] Ibid.

that 'a strong argument can be made that programs and specific curricula should be developed that involve animals with the aim of promoting moral awareness and fostering humane treatment of both human and non-human animals'.[69]

8. Something 'animal' in human morality

The role of emotions points to a dimension of morality that is not uniquely human. We are not the only animals that seem to be born with the disposition to have 'social' emotions.[70] The impulse to help others, for instance, is something we share with primates. Careful observations of primates reveal what could be called 'precursors' of moral reactions and behaviour.[71] Chimpanzees for example seem to have 'retributive emotions'.[72] They 'punish negative actions with other negative actions'.[73] These and many other observations support the 'long tradition, going back to Aristotle and Aquinas, which firmly anchors morality in the natural inclinations and desires of our species'.[74]

However, what can be found in our ancestors can only be called 'pre-moral' behaviour. The full development of the capacity for empathy in a human being requires a form of sentimental education that clearly lacks an analogue in the non-human animal world. In addition, we have some highly abstract cognitive capacities which make our morality 'unique in the animal kingdom'.[75] According to the psychologist Valerie Stone, these are the capacities for 'planning, inhibition, understanding others' mental states, and language'.[76] The ability to use language makes cognitive distancing possible and thus moral choices.[77] Stone argues that these 'high-level and abstract capacities', which primates don't have, enable humans to 'engage in a wide range of moral judgements and behaviours'.[78] While 'our ancient social instincts are the driving force of morality', the cognitive capacities concerned expand the range of behaviours which can be driven by those instincts.[79] This is Stone's modification of Darwin's claim that 'our moral sense depends on [our] intellectual capacities'.[80] If we didn't have

---

[69]   Ibid., 109.

[70]   It is actually contested whether empathy qualifies as an emotion. For the view that it does not, see, for example, Gary D. Sherman and Jonathan Haidt, 'Cuteness and Disgust: The Humanizing and Dehumanizing Effects of Emotion,' *Emotion Review*, 3 (2011), 247.

[71]   Frans de Waal, 'Morally Evolved: Primate Social Instincts, Human Morality, and the Rise and Fall of "Veneer Theory"', in *Primates and Philosophers: How Morality Evolved*, ed. Stephen Macedo and Josiah Ober (Princeton, NJ: Princeton University Press, 2006), 18f.

[72]   Ibid., 18. De Waal refers in particular to Edward Westermarck, *The Origin and Development of the Moral Ideas*, 2nd edn, 2 vols (London: Macmillan, 1912 and 1917).

[73]   De Waal, 'Morally Evolved: Primate Social Instincts, Human Morality, and the Rise and Fall of "Veneer Theory"', 18.

[74]   Ibid.

[75]   Valerie Stone, 'The Moral Dimensions of Human Social Intelligence: Domain-Specific and Domain-General Mechanisms,' *Philosophical Explorations*, 9 (2006), 56.

[76]   Ibid., 63.

[77]   Ibid., 56.

[78]   Ibid.

[79]   Ibid., 63.

[80]   Ibid., 62.

those capacities, we wouldn't play the variety of moral language-games we in fact play. Despite the trivial point that we wouldn't play any language-games if we didn't have the capacity for language, our moral language-games wouldn't be as various and elaborate if we didn't have the three other capacities mentioned by Stone. That humans have those highly abstract cognitive capacities can be said to belong to the constitutive 'surroundings' of moral language-games.

We can thus identify a further animalistic component of our morality: the emotions that are its driving force. Combined with the highly abstract cognitive capacities mentioned, our emotional capacities make us 'a moral animal'.[81] As de Waal argues, 'the human capacity to act well at least sometimes, rather than badly all the time, has its evolutionary origins in emotions that we share with other animals – in involuntary (unchosen, pre-rational) and physiologically obvious (thus observable) responses to the circumstances of others'.[82] However, the ways in which these emotions are educated in humans are 'something human'.

9. Conclusion

I have argued that the notion of empirical certainty with which Wittgenstein was concerned has a moral analogue. For competent players of our moral language-games, the ceteris paribus wrongness of killing, breaking a promise, lying, etc. is beyond reasonable doubt and at the same time not open to justification. This certainty is 'animal' in the sense that it is not the result of any reasoning process and is exhibited through unjustified ways of speaking and acting. In addition to critical reflection and deliberation, our moral language-games involve habitual ways of interacting with others as well as automatic emotional responses and spontaneous judgements.

As I have pointed out, the certainty of, for instance, the ceteris paribus wrongness of killing depends on moral competence: that, ceteris paribus, killing is wrong is certain for me *qua* morally competent agent. Moral learning is a complex process and involves the development of affective dispositions. I referred to research about animals as having a positive effect on moral development. Dealing with animals facilitates moral learning and fosters the development of empathy. The important role that empathy and other emotions have in partially constituting our moral competence places us closer, in certain respects, to some non-human animals. Although the ways in which our moral feelings are educated are particularly human, the emotional dimension of morality can be regarded as 'something animal'.[83]

---

[81] Ibid., 63. Since I take emotions to involve cognitive capacities, the contrast is not between emotional and cognitive capacities, but between emotional and highly abstract cognitive capacities.

[82] Josiah Ober and Stephen Macedo, 'Introduction', in *Primates and Philosophers: How Morality Evolved*, ed. Stephen Macedo and Josiah Ober (Princeton, NJ: Princeton University Press, 2006), xiii. As the editors pointed out to me, Ober and Macedo should have written 'amorally' instead of 'badly'.

[83] I thank the editors for their valuable comments and constructive criticism.

8

Living with Animals, Living as an Animal

Anne Le Goff

There is a whole bestiary to be found in Wittgenstein's writings. Grasshoppers, for example, appear in the *Remarks on the Philosophy of Psychology* (II). The way they are used is quite representative for the part played in general by animals in Wittgenstein's writings:

> 23. "Human beings think, grasshoppers don't." This means something like: the concept 'thinking' refers to human life, not to that of grasshoppers. And one could impart this to a person who doesn't understand the English word "thinking" and perhaps believes erroneously that it refers to something grasshoppers do.

> 24. "Grasshoppers don't think." Where does this belong? Is it an article of faith, or does it belong to natural history? If the latter, it ought to be a sentence something like: "Grasshoppers can't read and write." This sentence has a clear meaning, and even though it is perhaps never used, still it is easy to imagine a use for it.[1]

The point is not about grasshoppers themselves; rather, it is about human psychology and language. Grasshoppers play the role of a 'control group' that makes clear by contrast a feature possessed by human beings and not by grasshoppers, namely thinking. Animal forms of life in general constitute an 'object of comparison' that helps us distinguish features of our language-games, or of our own human 'form of life'.[2] 'Where does [the statement "Grasshoppers don't think"] belong' and what kind of knowledge is expressed by it? Wittgenstein suggests it might be pronounced as an article of faith – faith in a non-observable capacity of grasshoppers – but he does not endorse this thought. If it is to be a piece of natural history, we need to substitute for the original statement an observable version of it, such as: 'Grasshoppers can't read and write.' Though the sentence has a meaning and we can imagine a context where it could be used (e.g. said to a child who tries to teach her

---

[1] Ludwig Wittgenstein, *Remarks on the Philosophy of Psychology*, II (Oxford: Blackwell, 1980), §§23–4.

[2] However, the knowledge of the word 'thinking' gained by the present contrast is quite limited, for there are many things that human beings do and grasshoppers do not do (for instance, to run). The contrast does not define the word 'thinking'.

captured grasshopper how to read), it does not tell us much about grasshoppers and can hardly be considered as a fact of natural history since it could be applied to an infinity of things, more exactly to anything that is not *us*. This kind of statement about the animal does not bring any knowledge to bear on the animal, it merely places it in comparison to us.

Yet, it is not by chance that a grasshopper and not, say, a table is taken as a comparison point. The proposition that 'Human beings think, tables don't' is just as true as the original one. To state the difference with grasshoppers defines our concept of a human being in a way that stating it in relation to tables would not. While one can easily imagine a context where it would be relevant to compare human beings to tables, this comparison is not obviously meaningful. It seems that it is at least in a minimal way relevant to compare ourselves to grasshoppers. The reason why Wittgenstein uses animals for the purpose of comparison is that they share something with us beyond the mere fact of being alive (which is, of course, also a status shared by plants); that is, they possess *forms* of life that are at least somewhat similar to ours. It makes them appropriate for contrasting with, and bringing out specific features of the human form of life. Animals are in a sense our privileged others.

Cora Diamond and John McDowell, both profoundly influenced by Wittgenstein's work, recognize the importance of the idea that a human being is an animal. For both of them it entails a reflection on animal life in general and leads them to face the question of what philosophy can say about animal life. I want to take a closer look at the competing understandings of these questions that they offer. McDowell thinks that a human animal life is to be conceived as radically different from all other animal lives. Human life is the subject matter of philosophy, whereas animal life is not. For Diamond, the idea of human beings as animals means that we share a great deal with animals. To conceive of one implies, for philosophy, to conceive of the other.

In exploring these two perspectives, I will focus on what Diamond calls a 'difficulty of reality'.[3] Cases of extreme difficulty with reality reveal something of our ordinary way of being in the world, and particularly of our having a life. Diamond delineates the concept through a few literary examples and defines it in this way: such a difficulty arises in 'experiences in which we take something in reality to be resistant to our thinking it, or possibly to be painful in its inexplicability, difficult in that way, or perhaps awesome and astonishing in its inexplicability'.[4] In one of her examples, the difficulty directly concerns our relationship to animals. This example is the story *The Lives of Animals* by J. M. Coetzee.[5] The main character, the writer Elizabeth Costello, is horrified by the treatment we inflict on animals, to the point of being herself physically weakened.

---

[3] Cora Diamond, 'The Difficulty of Reality and the Difficulty of Philosophy,' in Stanley Cavell et al., *Philosophy and Animal Life* (New York: Columbia University Press, 2008). Diamond's essay was first published in *Partial Answers*, 1 (2003).

[4] Ibid., 45–6.

[5] J. M. Coetzee, *The Lives of Animals*, ed. Amy Gutmann (Princeton, NJ: Princeton University Press, 1999). Republished as two chapters of the novel *Elizabeth Costello* (New York: Viking, 2003). I quote from the first edition, used by Diamond. The philosophical apparatus around the story in this edition is also part of her example.

As in the other cases of difficulty of reality,[6] her ordinary sense of reality collapses in front of something she cannot acknowledge as reality; this breakthrough casts light on a part of her that tends to be concealed: her being alive as an animal. McDowell has written a comment on Diamond's essay.[7] It will be illuminating to consider both their accounts of this difficulty of reality, as it involves answering what it is to be a human animal or another animal. I will first briefly recall the main features of the example and how this difficulty of reality is also, according to Diamond, a difficulty of philosophy. At first sight, McDowell's understanding of the difficulty seems to match Diamond's. He develops a notion of the human animal as a rational or speaking animal. The crisis of one's ordinary relationship to reality results from one's inability to 'capture' reality in language. Yet, to underline this feature leads one to undermine other crucial aspects of what it is to be a human (or non-human) animal. I will argue that McDowell's treatment of the difficulty is but another case of what Diamond calls a 'deflection' of it: a deflection of our having a body, just like animals. Diamond opens another perspective for a philosophical conception of animal life in general through literature.

1. Costello's example

In J. M. Coetzee's story *The Lives of Animals*, Elizabeth Costello, a famous writer, is invited to give two lectures at the University where her son is employed. The lectures are on our ill-treatment of animals in contemporary society, and especially in the meat industry. Costello's way of tackling the issue – on the basis of literary texts but above all in a very personal way – is disconcerting and embarrassing to her audience. The hostile reaction is at its most intense when she draws a comparison with the Holocaust.[8]

What is this story about? In the first edition, the story was published with the addition of scholarly philosophical comments.[9] According to them, the story argues in favour of animal rights. Coetzee (who actually also read the stories as lectures) uses fiction to express arguments in order to vindicate a moral stance on a problem. The character of Costello is a literary device allowing him to embody a point of view that comes to expression during her lectures and her discussions with other characters. In that respect, Costello's own suffering could be interpreted as a rhetorical means of

---

[6] I will not be able to broach here the other cases of difficulty of reality. Diamond takes as examples a poem by Ted Hughes that brings a sense of death at the core of life; absolute beauty; inexplicable goodness, in Ruth Klüger's memoirs; a story by Mary Mann that brings 'spikiness with morals'; Cavell's discussion of *The Winter's Tale* and *Othello* with respect to scepticism and knowledge of the other in his *The Claim of Reason* (Oxford: Oxford University Press, 1979), 481–96. Though Coetzee's story shows it the most clearly, all these cases are related to the fact of re-discovering one's being alive and the limitations it induces.

[7] John McDowell, 'Comment on Stanley Cavell's "Companionable Thinking"', in *Philosophy and Animal Life*. As the title shows, McDowell's essay is actually a response to Cavell's essay on Diamond: Stanley Cavell, 'Companionable Thinking', in *Philosophy and Animal Life*. Cavell's and McDowell's essays were first published in *Wittgenstein and the Moral Life: Essays in Honor of Cora Diamond*, ed. Alice Crary (Cambridge, MA: MIT Press, 2007).

[8] Coetzee, *The Lives of Animals*, 19–22.

[9] Comments by Amy Gutmann, Marjorie Garber, Peter Singer, Wendy Doniger and Barbara Smuts.

strengthening the argument. From this point of view, fiction is reduced to a 'cloth[ing]'[10] of ideas or arguments. Yet, such an interpretation does not make much sense from a literary point of view. Is *The Lives of Animals* simply part of an ideological novel like Voltaire's *Candide* or Diderot's *La Religieuse*? There is more to this story than an argumentative content, notably its complex characters (first of all, Elizabeth Costello) who are not reducible to mere mouthpieces for philosophical claims. Furthermore, it is not clear that Costello is providing arguments. Or if she tries to do so, she does not do a very good job of it. This becomes especially clear in her ambiguous use of the Holocaust comparison: though she thinks it allows her to 'score', it attracts a great deal of hostility towards her. Even though she expected this reaction, she does not manage to make the others see the argumentative value of her point.[11] What we can see if we pay attention to the story itself is that Costello offers something: she lets her audience see *her*, as she is onstage, or rather she hands herself over to her audience, her person and her suffering *for* animals and (as Diamond points out) *as* an animal. While, according to certain academic conventions, the particular subjectivity of the speaker is expected to disappear behind the objective argument, Costello does not conceal herself behind the stories. She takes an emotional part in them. In the same way, the narrator lets us see her: he depicts her elderly body, the evident weariness of it.[12] In Diamond's words, Coetzee's lectures present 'a kind of woundedness or hauntedness, a terrible rawness of nerves'.[13] Costello is wounded by the perpetual wound we inflict on animals. As the notion of 'wound' makes clear, the difficulty is not only intellectual but it also affects her body. As the term 'haunted' suggests, it is not a problem she could put aside but it looms over her entire life.

What makes the wound terrible is her isolation: other people do not seem to feel the horror. Indeed, they can (or at least they think they can) deal argumentatively with it. As Diamond puts it, 'one thing that wounds her is precisely the common and taken-for-granted mode of thought that "how we should treat animals" is an "ethical issue" '.[14] But to regard it as an ethical issue that can be argued for implies a distance between us, who reflect on and judge the issue, and them, the objects. The only possible response for Costello is, on the contrary, the absence of distance: the suffering with the animals as an animal herself. Though deeply ethical also, her words and attitudes do not constitute a response to an 'ethical issue'. She rejects the argumentative discussion, regarding it as a way of avoiding the only correct response. This point comes sharply into view in her use of the analogy of the Holocaust. She embraces the inevitable consequence of being rejected by her interlocutors, of being isolated.

---

[10]   Diamond, 'The Difficulty of Reality and the Difficulty of Philosophy', 53.
[11]   Coetzee, *The Lives of Animals*, 22: 'Pardon me. I repeat. This is the last cheap point I will be scoring.' Why is it a cheap point? Not because the argument is dubious, rather because her choice of weapon is not quite fair: it is a weapon her opponents do not agree on using. Still, she insists on using it for it is a valid, though unpleasant, argument.
[12]   See, for instance, the initial description of her in Coetzee, *The Lives of Animals*, 16.
[13]   Diamond, 'The Difficulty of Reality and the Difficulty of Philosophy', 47.
[14]   Ibid., 51.

As Diamond shows, the commentators on Coetzee's story miss the point just as Costello's audience (Norma, Leahy . . .) miss it, building a sort of *mise en abyme* (mirroring effect). Inside the story, most of her interlocutors take Costello to be propounding arguments in an attempt to defend animal rights, while in fact she is doing something else. Similarly, commentators have generally taken Coetzee to be giving arguments, whereas *he too* is doing something else. Diamond describes these common forms of interpretation by invoking the Cavellian concept of deflection.[15] Deflection is 'what happens when we are moved from the appreciation, or attempt at appreciation, of a difficulty of reality to a philosophical or moral problem apparently in the vicinity'.[16] The difficulty exemplified by Costello is not about an ethical problem that can be solved by taking into account the various requirements of all parties. It obtains at a much deeper level: it is not that something *in* reality is problematic (such as its being, or not being, morally wrong to eat animals); rather, reality itself has become a problem – our own existence has become a problem to us. To be deflected from the real difficulty means here not to take into account that we also have a body, that we also are animals. Commentators abstract themselves from their lives to reflect on an abstract question. By bringing the problem to a certain philosophical level, they avoid the real difficulty that is brought into view by the story, generically defined by Diamond as 'the experience of the mind's not being able to encompass something which it encounters'.[17] The question is: in what exactly does the difficulty consist? Addressing this question will negatively provide an insight into our ordinary way of living and thinking. Let us examine McDowell's account of this difficulty through his notion of the human being as a rational animal.

2. Speaking animals and 'mere' animals

Human beings as rational animals

In the cases of difficulty of reality, McDowell glosses, 'something we encounter defeats our ordinary capacity to get our minds around reality, that is, our capacity to capture reality in language'.[18] The reason why these difficulties are so devastating is that language is our specific mode of being in the world. Reality is given to us in a linguistic shape. A difficulty of reality, in its resistance to our expressing it, puts in question our very existence as 'speaking animals', '[t]he special kind of animal life we lead'.[19] It is not only a partial failure; rather, our whole grip on reality – and, as a consequence, our conception of ourselves – is shattered. In another context, McDowell has expounded on this conception of the human being by means of the Aristotelian concept of a

---

15 Stanley Cavell, 'Knowing and Acknowledging,' in *Must We Mean What We Say?* (New York: Scribner's, 1969), 247, 260.
16 Diamond, 'The Difficulty of Reality and the Difficulty of Philosophy,' 57.
17 Ibid., 44.
18 McDowell, 'Comment on Stanley Cavell's "Companionable Thinking",' 134.
19 Ibid.

rational animal.[20] His idea is that becoming rational is 'our way of actualizing ourselves as animals'.[21] Our way of being animal is to be rational. Exercises of reason (or of 'spontaneity', in McDowell's Kantian parlance) 'belong to our mode of living'.[22] To put it differently, reason is a 'second nature' to human beings. This idea of second nature aims to show how to understand rational capacities as autonomous (i.e. non-reducible to physical or biological properties) and as natural: these capacities are natural in the sense that we have acquired them through our education without the intervention of any other (supernatural) factor. Language is the human being's way of existing, just as the beaver's own way of existing is to live in a semi-aquatic environment where, among other things, it builds dams. McDowell himself offers this analogy.[23] His point is not to claim that the beaver is like the human being. Quite the opposite; it is to show that the human being is like the beaver, in that she is also part of nature and inhabits the world in a specific way. This human way of inhabiting the world is language. The beaver acquires certain practices and abilities in response to the demands of its environment and in interaction with its congeners. In the same way, the human (rational) environment (with fellow humans) exerts a determining influence on the child's development. At the same time as the child learns how to behave in this world, she learns how to talk and think.

The examples identified by Diamond as 'difficulties of reality' bring to view, in a negative light, that language and reason are our second nature. Costello lost her ordinary grip on reality through language. This is clear in the dialogue with her son at the very end of the text. Her son says he has not 'had time to make sense of why [she has] become so intense about the animal business'. Costello replies: 'A better explanation . . . is that I have not told you why, or dare not tell you. When I think of the words, they seem so outrageous that they are best spoken into a pillow or into a hole in the ground like King Midas'.[24] Her familiar words, her life-long allies, have become 'outrageous', unrecognizable. What she sees and tells is the horrible way in which we daily treat 'our fellow creatures' (as Diamond puts it in another context).[25] But, obviously, no one else understands the words she pronounces: she does not speak the same language as the others. Her own son does not understand her in this last discussion, as his choice of neutral words such as 'the animal business' proves. They do not talk about the same thing. As Costello loses common language, the world collapses for her – the only possible world, the one of shared meanings. She faces the sort of difficulty of reality that, as McDowell puts it, 'dislodges us from comfortably inhabiting our nature as speaking animals'.[26] This is all the more true of her, being an aged writer: her familiar world is language. She is no better than the king made donkey who, despite his looks,

[20] In John McDowell, *Mind and World* (Cambridge, MA: Harvard University Press, 1996), Lectures IV and V, and in several subsequent articles and responses.
[21] McDowell, *Mind and World*, 78.
[22] Ibid.
[23] McDowell, 'Comment on Stanley Cavell's "Companionable Thinking"', 134.
[24] Coetzee, *The Lives of Animals*, 69.
[25] Cora Diamond, 'Eating Meat and Eating People,' in her *The Realistic Spirit* (Cambridge, MA: MIT Press, 1991), 319–34, at 328f.
[26] McDowell, 'Comment on Stanley Cavell's "Companionable Thinking"', 134.

lost his own humanity. Costello is 'unhinged', says McDowell.[27] In this word, we can hear Wittgenstein's concept of hinge propositions.[28] Hinge propositions are the certainties around which all our other beliefs revolve, precisely because they remain fixed and never questioned. Costello's most intimate certainties collapse as she witnesses 'a crime of stupefying proportions' being committed with everyone else's agreement.[29] All that is left to her is to note with despair: 'I no longer know where I am.'[30]

Difference from other animals

McDowell's use of the idea of second nature allows us to understand our possessing *logos* (i.e. both reason and language) as well as our being animals. Yet, if we are animals, it has to be in a very different sense from what he calls 'mere' animals. Our specific (rational) capacities, though perfectly natural, dramatically distinguish our mode of life from the one of other animals. We need to resist the temptation to assume that there is a continuity between two kinds of animality, human animality and (so to speak) animal animality, a temptation voiced by John (Elizabeth Costello's son): 'isn't there a position outside from which our doing our thinking and then sending out a Mars probe looks a lot like a squirrel doing its thinking and then dashing out and snatching a nut?'[31] John's wife Norma, a philosopher, replies that such an idea, to the effect that 'rational accounts are merely a consequence of the structure of the human mind . . . is shallow relativism that impresses freshmen'.[32] This also is McDowell's stance: we need to resist jumping to identities from such similarities. The point of the analogy between the human being and the beaver needs to be carefully dissected: the analogy shows that it is as natural (in the full sense of the term) for the human being to talk and think as it is for the beaver to build dams. It does not show that reason or linguistic capacities are *of the same nature* as the beaver's dam-building abilities. The beaver is an animal adapted to its environment; the dam-building technique aims to provide a solution to a precise issue. Language use is a wholly different ability that lets us access a wholly different space, the 'space of reasons' or meaning.[33] McDowell makes use of a distinction he finds in Hans-Georg Gadamer: submission to an environment versus orientation to the world.[34] Human beings differ from other animals because their education, instead of giving them control over their environment, opens the world to them. The world is the objective reality, while the environment is merely a source of satisfaction of subjective needs. '[M]erely animal life is shaped by goals

---

[27] Ibid., 136. Diamond first uses this term in connection with Hughes' poem, in 'The Difficulty of Reality and the Difficulty of Philosophy', 58.

[28] See Ludwig Wittgenstein, *On Certainty* (New York: Harper and Row, 1972), §§341–6, 150–3.

[29] Coetzee, *The Lives of Animals*, 69.

[30] Ibid.

[31] Ibid., 48.

[32] Ibid.

[33] McDowell borrows the concept of the 'space of reasons' from Wilfrid Sellars; see McDowell, *Mind and World*, 5.

[34] Hans-Georg Gadamer, *Truth and Method* (New York: Crossroads, 1992), 438–56; in *Mind and World*, 115–19.

whose control of the animal's behaviour at a given moment is the immediate outcome of biological forces.'[35] Only rational human beings act under reasons and can thus be free of 'enslavement to immediate biological imperatives'.[36] Of course, McDowell does not pretend that human actions would always result from a conscious weighing of reasons. But the point is that human beings can give a justification for their actions if required, while mere animals cannot.

Our having such different kinds of life means that we do not share anything with animals that could be understood as a 'highest common factor'.[37] McDowell elaborates on this claim in the case of perception. While he acknowledges that we share with animals perception or a 'perceptual sensitivity to our environment',[38] he denies that we possess it in the same sense. As our perception 'is taken into the ambit of the faculty of spontaneity', we have it 'in a special form', – that is, a conceptual form.[39] The content of our perception is already conceptual, whereas it cannot be for animals since they do not have concepts. He gives the following example: a cat and a human being find on their way a wall with a hole in it.[40] Both will view the hole and both will be able to use it to go through to the other side of the wall. However, although it is the same thing that the cat and the human being perceive, their perceptions of it differ. The description according to which they perceive the hole is true for both only at a superficial level. The cat's perception and my perception are only homonyms. The same concepts (of hole, wall, etc.) are not in play in both of the descriptions that we may give. In such a case, '[m]y experience [as a human being] would be world-disclosing and so conceptual in form', while '[t]he cat's perceptual intake' would not be an experience of the world, only a mere response to an affordance.[41]

Accordingly, philosophical remarks about animals can only be of the type exemplified by Wittgenstein's remark on grasshoppers: negative remarks in comparison to us. This is all that McDowell offers (animals don't have a world, they don't perceive or act in the full sense of these terms, etc.). He concedes that concepts like 'orientation' or 'proto-subjectivity' can be useful to make sense of animal behaviour, but only in cognitive science and not in philosophy, where they should be reserved for describing the life of the beings who can access reasons.[42] Reality is divided into two realms: the realm of so-called first nature, defined as the domain of the natural sciences, and the realm of reasons. As animals are fully contained within the so-called first nature, they are apt to be studied by certain of the natural sciences (such as zoology, cognitive ethology and comparative psychology). Philosophy, meanwhile, is competent only to investigate second nature.

[35] McDowell, *Mind and World*, 115.
[36] Ibid., 117.
[37] Ibid., 113. This supposed 'highest common factor' is the 'experiential content of a kind we share with mere animals', that is 'non-conceptual' content (ibid., 114).
[38] Ibid., 64.
[39] Ibid.
[40] John McDowell, 'What Myth?' in *The Engaged Intellect: Philosophical Essays* (Cambridge, MA: Harvard University Press, 2009), 314f.
[41] Ibid., 321.
[42] See McDowell, *Mind and World*, 121.

Yet, this radical gap among living beings sounds unfair both to animals and to human beings. It is first to be noticed – though I will not develop this line of criticism here – that McDowell's distinction is based on a very questionable picture of animal life as enslaved to life pressure, and hence is hardly able to account for the creativity and intelligence displayed in many animal behaviours. Moreover, to identify our specific mode of living with our possessing language as McDowell tends to do is itself questionable. I will develop this second line of criticism below.

3. Living animals

Bodies

McDowell is right to underline that at the core of our contingent, animal existence there is language. However, he tends to go beyond this fruitful thesis to claim that exercises of reason *make up* our mode of living. This claim is made at the expense of acknowledging what we might share with animals. According to him, our being animal is precisely our *not* being animal in the sense of the other animals. This picture of human life leads him to overlook the complexity of the cases of difficulty of reality. The difficulty faced by Costello is not only an intellectual but also a physical experience. McDowell agrees to a certain extent when he writes: 'For Costello, it becomes a problem to live her particular case of the lives of animals: a life in which words are not just a distinguishing mark, as they are for human animals in general, but the central element. Her being as the animal she is, which is her bodily being, becomes a wound.'[43] On the face of it, McDowell agrees with Diamond: Costello does not only suffer *for* the animals but also herself *as* an animal. But what does the wound consist of according to him? The wound is inflicted upon her as a speaking being, and is all the more severe for Costello who is a writer, an extreme form of a language-using animal. It is because the wound touches her at the core that her being as a whole is affected. In McDowell's reading, it is only because she *also* is an animal that Costello consequently and secondarily is hit in her body and life. She suffers as an animal, not because she shares something with the other animals, but, on the contrary, because she differs from them and suffers as a linguistic being.

This is not what Coetzee's text says. Costello's wound is not a consequence of her linguistic incapacity. The wound is double-edged: it encompasses her whole being in the world, both as bodily and linguistic – her whole life. As Diamond remarks, Elizabeth Costello is a wounded animal *herself*. 'She describes herself as an animal exhibiting but not exhibiting, to a gathering of scholars, a wound.'[44] The text repeatedly lays bare her animality and her non-metaphorical wound. For instance, the initial description of her only mentions her being 'fleshy' and 'white-haired'.[45] The text shows us a woman who looks more and more fragile as the story goes on, 'the woman with

43   McDowell, 'Comment to Stanley Cavell's "Companionable Thinking"', 134.
44   Diamond, 'The Difficulty of Reality and the Difficulty of Philosophy', 47.
45   Coetzee, *The Lives of Animals*, 16.

the haunted mind and the raw nerves'.[46] In the end, she cannot but collapse in tears. To be sure, part of her suffering is her inability to find herself in language. But this inability is inseparable from her vulnerability as an animal. A strong moment in her lecture is when she says:

> For instants at a time . . . I know what it is like to be a corpse. The knowledge repels me. It fills me with terror; I shy away from it, I refuse to entertain it. . . . The knowledge we have is not abstract: 'All humans are mortal, I am a human being, therefore I am mortal' – but embodied. For a moment we *are* that knowledge.[47]

Even though words are failing her, she has a knowledge and a particularly clear one. It has the devastating clarity of pain. It is too hard to examine it by thought, words are weak; but she can undergo it – more exactly, she cannot help undergoing it, to the point of *being* this knowledge. Instead of mastering her knowledge, she is possessed by it, as by a demon that annihilates everything else in her.

Deflection

McDowell, though he would of course not deny that their having a body is important to the existence of human beings, regards it a mere fact. This is only a contingent fact because it does not define us; what defines us is our reason. Though our bodily makeup imposes some constraints and limits on our perception and action in the world, it does not contribute to defining them. Our perceptual and active capacities are appropriately described as rational. Accordingly, Costello's bodily pain is merely a collateral consequence of her existential suffering, not a part of it. But to make 'our own bodies mere facts', Diamond claims, is a deflection from the real difficulty into a well-delimited philosophical problem.[48] Whereas McDowell claims that where we are 'at home' is 'in the space of reasons',[49] 'Coetzee's lectures ask us to inhabit a body'.[50] Our body is one (not the only one) of the places where the never-ending debate between the animal and its world is at play. We cannot understand what is at stake with the difficulties of reality if we do not take the body into account. Diamond underlines 'how much this coming apart of thought and reality belongs to flesh and blood'.[51] If the life and death of animals can 'unhinge' our reason, it is not as facts that will be judged relevant or not, but as presences one cannot escape.

The philosophers criticized by Diamond are deflected from the real difficulty by moving it to the level of ethical issues and attribution of rights. They make the animals they are concerned with into abstract 'living beings', rather than real-life pigs, apes or pandas. Each abstract living being possesses sentience (in Singer's Utilitarian

---

[46] Diamond, 'The Difficulty of Reality and the Difficulty of Philosophy', 48.
[47] Coetzee, *The Lives of Animals*, 32.
[48] Diamond, 'The Difficulty of Reality and the Difficulty of Philosophy', 59.
[49] McDowell, *Mind and World*, 125, *et passim*.
[50] Diamond, 'The Difficulty of Reality and the Difficulty of Philosophy', 59.
[51] Ibid., 78.

perspective)[52] or is a 'subject-of-a-life' (in Regan's Kantian perspective).[53] McDowell is deflected in another way: he is deflected from the questions raised by animals (about our relationship to them and about ourselves) by problems in the philosophy of perception and action. The only issue for him is whether we should grant perception and action to animals in the same sense as that in which we grant them to ourselves. As his answer is negative, philosophy's task is to ground the claim of the anthropological difference. Yet, McDowell would not acknowledge that he is deflected. He would simply reply that he and Diamond deal with different philosophical issues and that our having a body in the sense emphasized by Diamond, and our being animals as such, are not *his* problems. Animal life is not a topic of its own in his work, it emerges only as a possibly problematic consequence of his philosophy of perception. But this is exactly what Diamond describes as 'deflection': the deflected problem does not come into view at all as something relevant and that needs to be faced. McDowell's case is all the more interesting because, contrary to many philosophers, he does pay attention to our animality and strives to take it into account. Yet, he does not manage to grasp the full extent of it.

Exposure

What we are deflected from is, in a general sense, our 'exposure'. Diamond borrows this concept from Cavell.[54] 'Being exposed', regarding a concept, means that 'my assurance in applying the concept isn't provided for me'.[55] There is no authority to which we can appeal to guarantee that we are applying it correctly, besides that of our own judgement. McDowell's idea that we are at home in the space of reasons is our ordinary illusion. The cases of difficulty of reality reveal how this 'being at home' is precarious. They are a 'repudiation of the everyday'.[56] The failure of our common ways of thinking and talking reveal their deep and usually invisible fragility. We are always at risk of losing our (bodily as well as linguistic) grip on reality.

Diamond applies Cavell's concept to the case of one's relationship to animals. Circumstances entitle me to treat a certain concept of animal as relevant rather than another one (rabbit as a pet or rabbit that is raised for food). Nothing or no one else can guarantee that I am applying the correct concept of animal in the particular situation I am in. No general principle is available to me, nor are any essential properties to be found in the animals that would *tell* me what to do. This leaves us with only our own responsibility to act in the appropriate way, 'our own making the best of it'.[57] It does not mean, however, that nothing is justifiable and therefore everything is permitted. The point is that there is no justification outside of the particular non-ideal context, which means there is no ultimate justification. It also entails that our attitudes are bound not

[52] See Peter Singer, *Animal Liberation* (New York: Random House, 1975).
[53] See Tom Regan, *The Case for Animal Rights* (Berkeley, CA: University of California Press, 1983).
[54] Stanley Cavell, *The Claim of Reason* (Oxford: Oxford University Press, 1979), 433, 439.
[55] Diamond, 'The Difficulty of Reality and the Difficulty of Philosophy', 71.
[56] Ibid.
[57] Ibid., 72.

to be fully consistent: 'it may at best be a kind of bitter-tasting compromise'.[58] Coetzee's Elizabeth Costello is an example of someone who acknowledges her exposure. She feels very strongly – even too strongly for her own well-being – her responsibility towards animals as she explicitly expresses in her lectures and personal discussions, and implicitly in her whole attitude. Nevertheless, her awareness does not allow her to fully escape her exposure; even she compromises: as she herself points out, she carries a leather purse and wears leather shoes.[59]

There is also what I would call a second-order exposure, felt by Costello and made reflectively obvious by the difficulty posed by our relationship to animals: namely, our having bodies. Our exposure lies in the fact that our animal life is by definition vulnerable. We discover it when 'we find ourselves . . . in a shuddering awareness of death and life together'.[60] As Diamond puts it: 'The awareness we have of being a living body, being "alive to the world", carries with it exposure to the bodily sense of vulnerability to death, sheer animal vulnerability, the vulnerability we share with them.'[61] This exposure lies at the bottom of the difficulty of our relationship to animals. Costello's special difficulty does not arise only from a failure of understanding or conceiving (as McDowell implies), nor from a sense of her absolute responsibility in defining who should die or live (as Diamond's above definition of the concept of exposure suggests), but rather from her being *both* the animal that kills and the animal that is killed. In this connection, a passage from which I quoted earlier is relevant:

> "For instants at a time," his [i.e., John's] mother is saying, "I know what it is like to be a corpse. The knowledge repels me. It fills me with terror; I shy away from it, refuse to entertain it.
>
> All of us have such moments, particularly as we grow older. . . . For a moment we *are* that knowledge. We live the impossible: we live beyond our death, look back on it, yet look back as only a dead self can."[62]

Knowledge of death in the strong sense is not factual knowledge of something that befalls people we know or love and that will befall us one day. It is knowledge of one's death as a premonitory experience of death in life. From these instants, Costello[63] has gained knowledge of what it is to be a living being that is dying, like the cattle in the slaughterhouse. In the same way, she says about Hughes' poem, 'The Jaguar':[64] 'When we read the jaguar poem, when we recollect it afterwards in tranquillity, we are for a brief while the jaguar. He ripples within us, he takes over our body, he is us.'[65] Literature

[58] Ibid.
[59] Coetzee, *The Lives of Animals*, 43.
[60] Diamond, 'The Difficulty of Reality and the Difficulty of Philosophy', 73.
[61] Ibid., 74.
[62] Coetzee, *The Lives of Animals*, 32.
[63] She is here designated, not by her name, but by 'his mother'. The mention of her relation to her son highlights her being a living being, in lieu of her role as a great writer and speaker, a mind.
[64] Ted Hughes, 'The Jaguar', in *Ravens* (London: Rainbow Press, 1979), cited in *The Lives of Animals*, 50f.
[65] Coetzee, *The Lives of Animals*, 53.

allows one to experience another being's life, in this case the imprisoned one. But her tragedy – 'difficulty' seems too weak a word – is that she also is the perpetrator of their death or ill-treatment inasmuch as she belongs to the community of human beings. She is both the killed pig and the killer; she stands inside the jaguar cage and outside it.[66] As Diamond puts it, 'she describes herself as an animal exhibiting but not exhibiting . . . a wound which her clothes cover up, but which is touched on in every word she speaks'.[67] Though she feels a kinship with all these suffering animals, she cannot deny she belongs to the human community that is the author of their ills. She accepts it, putting on clothes to conceal her body. Her language is by definition a human language; it fails to articulate what she means to say and yet she has no other way of expressing herself.

Can philosophy say anything about animals?

This exposure, so vividly exemplified in the character of Elizabeth Costello and in the difficulties of reality in general, is something difficult to contemplate, let alone to think through. What if, as Diamond puts it, philosophy 'does not know how to treat a wounded body as anything but a fact[?]'[68] It seems that philosophy, as a production and use of concepts, will only miss what is 'flesh and blood',[69] emotions and feelings: that is, not only the wounded body, but the body per se. In that case, there would indeed be nothing more for philosophy to say about animals than something like 'Grasshoppers don't think.' Not because to say something more would be the task of cognitive science, as McDowell suggests, but because what is to be thought slips out of the hands of philosophy.

Diamond is well aware of this difficulty of philosophy and she gives Simone Weil as an example of a philosopher who maintained an awareness of something that should not be treated as a fact, namely affliction.[70] Diamond herself gives another example of how philosophy can gain this awareness through literature. Writers and poets let us 'inhabit' the body of 'an imagined other',[71] and such a feat is particularly relevant in the case of animals. Costello tries to let her audience come into the body of the jaguar through Hughes' poem or into the body of Kafka's Red Peter. Coetzee invites us to be in another animal's body, namely Costello's. Yet, the purpose of literary imagination is not to fully identify with the animal in question, not with Costello and even less with the other non-human animals. In Diamond's words, we have 'a sense of astonishment and incomprehension that there should be beings so like us, so unlike us'.[72] This sense of astonishment is also part of the difficulty of reality concerning animals. McDowell

---

[66]   Guard and onlooker, a standpoint brought to view by another poem she mentions, Rilke's 'The Panther.' See Rainer Maria Rilke, 'The Panther,' in *New Poems* (San Francisco, CA: North Point Press, 2001), mentioned in *The Lives of Animals*, 50.

[67]   Diamond, 'The Difficulty of Reality and the Difficulty of Philosophy,' 47.

[68]   Ibid., 59.

[69]   Ibid., 78.

[70]   Ibid., 74–6. See Simone Weil, *Simone Weil: An Anthology*, ed. Siân Miles (New York: Weidenfeld and Nicholson, 1986).

[71]   Diamond, 'The Difficulty of Reality and the Difficulty of Philosophy,' 59.

[72]   Ibid., 61.

(for instance), in insisting on a radical difference between speaking human beings and merely biological animals, strives to remove this sense of astonishment. Literature, on the other hand, does not deny this sense of astonishment; rather, it helps us cultivate it. When the jaguar 'ripples within us', he comes for a brief while, thanks to poetry, and then goes. We do not entirely become the jaguar. Literature allows us to move in the in-between: to enter someone else's emotions, hopes, fears, etc. – almost to be someone else – for a little while. This idea of 'beings so like us, so unlike us' is implicit in Wittgenstein's remark about grasshoppers. The remark highlights the radical difference between two forms of life – one that involves thought and the other that does not. But this difference takes place against the background of a deep similarity – namely, the having of a life. While Wittgenstein's observation is bound to be merely negative and non-informative about grasshoppers, literature offers the opportunity for positive statements besides those made by biology or ethology. Such statements need a context in which to be made, literature provides it.

Let us consider as an example the poem 'Titmouse' by Walter de la Mare, commented on by Diamond.[73] The poet finds in the bird a 'happy company'. He has made a birdhouse for the titmouse to 'take his commons there'. They both share housing ('commons') and, in the etymological meaning of 'company', break bread together. The bird takes part in these basic features of social life. As Diamond puts it, the titmouse is a 'fellow creature' for the poet. He calls him 'this tiny son of life', bringing into view their kinship, the kinship of two sons of life in different guises. It will probably be objected that such characterizations of the bird are nothing but anthropomorphic. It may seem to some that by referring to a titmouse as 'company' we are merely projecting a feature of our form of life onto it. It does not bring us any further from ourselves nor closer to the animal than 'grasshoppers don't think'. But this critique proceeds from a view similar to the one implicit in McDowell, a view that the only legitimate perspective on animals is a natural-scientific, biological, one. Yet, Diamond remarks: 'it is not a *fact* that a titmouse *has a life*; if one speaks that way it expresses a particular relation within a broadly specifiable range to titmice. It is no more biological than it would be a biological point should you call another person a "traveller between life and death": that is not a biological point dressed up in poetical language.'[74] It is clear that the concept of life with respect to human beings is not merely (nor mainly) a biological one. In most of our usual practices towards animals and ways of considering them we do not treat them as mere organisms. This conception of animal life as biological actually begs the question. An illustration of this prejudice was given by ethology. Jane Goodall, a pioneer of field ethology, broke one of ethology's long-standing customs by giving names instead of numbers to the Gombe chimpanzees she was observing.[75] She knew she would be accused of anthropomorphism and sentimentality, and of engaging in 'non-scientific' behaviour. It should first be noticed that it is probably quite illusory to believe that an observer would feel nothing for the individuals she or he observes,

---

[73] Diamond, 'Eating Meat and Eating People,' 328f.
[74] Ibid., 330.
[75] See Jane Goodall, *In the Shadow of Man* (Boston, MA: Houghton Mifflin, 1988), 32.

but this is not what she said in her defence. She said that it was easier for her work. It is easier, indeed, because the chimpanzees are individuals, and not merely specimens of a species. In order to observe their personalities and relationships with each other, she needed to be able to recognize them. Nothing prevents us from using non-biological concepts to conceive of animal life, such as 'company', and it even seems necessary sometimes to use such concepts to accurately describe animal life. To call the titmouse a 'son of life' is not merely to use a metaphor. It expresses a kinship between the human being and the animal, such that there is something one can share with a bird or another animal that one cannot share with a piece of furniture of a rock. The bird is not only there, he is alive and unpredictable; this is why he can be company.

Though it truly is a challenge for philosophy to conceptually grasp animal life without distorting its meaning, it is both necessary and possible to try. Necessary, first, because even when philosophy does not mean to say anything about animals, it cannot help doing so in connection to human life. This is what McDowell's example reveals. No full understanding of human life can be achieved without taking into account what human beings share with animals. Secondly, Diamond has shown that a philosophical thinking of animals is not impossible, provided philosophy makes good use of other perspectives on animals. Literature, in particular, offers a very fertile resource.[76]

[76] I warmly thank the editors for their very helpful input.

9

What's Wrong with a Bite of Dog?

Rami Gudovitch

To many of us, the mere thought of eating a cat or a dog is outrageous; however, we can easily imagine expressing our outrage at a dinner conversation, over a burger. Does this gap between our attitudes to cats and dogs versus our attitudes to cows betray a deep inconsistency or hypocrisy on our part? This essay argues that it does not.

1. Approaches to morality

Morality is often understood as resting on an impartial approach towards moral subjects.[1] Accordingly, it is assumed that the ultimate justification of our moral principles must not rest on our likes, dislikes or other personal biases. The image of impartiality seems to stand at the very heart of the idea of morality – isn't the equal and fair applicability of moral principles to all just what morality is about? One challenge to this conception of morality is raised by the Wittgensteinian school, or at least, by some of its recent advocates. Authors such as Cora Diamond and John McDowell renounce the image of 'impartiality' as standing at the core of morality.[2] The aim of the present essay, along the lines of this new Wittgensteinian tradition, is to argue that human attitudes play an inextricable role in any viable conception of morality.

The proposed argument concentrates on the question of the moral status of animals. In particular, it defends the tenability of a moral view involving preferring some animals to others. But the purpose of defending the coherence of a preferentialist view towards animals is not to argue for the legitimacy of a preferential attitude to different individual humans, communities or ethnic groups. Rather, the purpose is to highlight the inextricable role our attitudes and preferences play in any form of moral reasoning.

---

[1] By 'impartial' I mean that, ultimately, moral justification turns to principles or rules that are general, and do not involve in their specification any reference to their holder, her attitudes or the specific characteristics of the situation involved.

[2] See John McDowell, *Mind and World* (Cambridge, MA: Harvard University Press, 1994), 34–6; Cora Diamond, *The Realistic Spirit* (Cambridge, MA: MIT Press, 1991), ch. 15.

2. Attitudes to animals

Animals play a complex role in our lives. Our attitudes towards animals are full of tensions and contradictions. Different individuals and communities share some of their attitudes to animals whereas they differ in others. Many of us share many of our attitudes to farm animals, pests and pets, for example. However, some vegetarians describe the meat industry in terms commonly reserved to describe the worst human atrocities, while many carnivores simply fail to see what the fuss is about.

Such differences in people's attitudes to animals are often correlated with substantial differences in their conception of the moral status of animals. An impartialist conception of morality requires that such moral conflicts be settled on the basis of some robust factual grounds, discernible independently of one's attitudes and biases, by virtue of which a subject (human or otherwise) merits moral attention.

There are different conceptions in the philosophical literature regarding the types of facts that could serve for grounding a creature's merit for moral attention. Among the candidates are facts about the creature's sentience, its cognitive skills or its degree of rationality. But obvious counterexamples suggest that none of these candidates does justice even to our most basic moral intuitions. Octopi are claimed to have a richer set of neural sensors than dogs, and therefore are likely to be capable of suffering more extreme forms of pain; some claim that a computer has a higher degree of cognitive complexity than a cat, while some chimpanzees express more rationality in their behaviour than some unfortunate humans. Still, most of us are unlikely to find the corollary claims – for example, that chimpanzees' claim for moral attention is stronger than that of some unfortunate human beings – acceptable; such claims contradict some of our firmest beliefs about humans and their moral status.

The impartialist, on the other hand, does not find such results objectionable. She argues that since the dictates of a solid and valid moral reasoning might not coincide with all our pre-theoretical 'moral' intuitions, such conflicts are not surprising. On the contrary, in cases where such conflicts occur, we must overcome our pre-theoretical intuitions and change our judgements so that they would accord with the valid dictates of morality. The Utilitarian Peter Singer, for example, in some of his writings, defends a view according to which the commands of morality stem from the demand to respect other creatures' interests. Facts about a creature's interests, for Singer, are objective, and are independent of the concerns and attitudes others have about the creature in question, its welfare, etc. Some interests are shared by all creatures, such as the interest to avoid suffering, while other interests depend on the creature's capacities and potentialities. According to this position, moral principles and judgements must be indifferent to the concerns we happen to have for the interests of some but not of others, and concentrates on facts about interests alone.[3]

The Wittgensteinian objects to the impartialist proposal, arguing that morality must not be divorced – not even locally and in the interest of systemization – from the

---

[3] See, for example, Peter Singer, *Practical Ethics* (Cambridge: Cambridge University Press, 1993), esp. 21.

complex network of considerations, motivations, judgements and emotions regarding who merits moral attention and when.[4]

The Wittgensteinian and the impartialist can agree that there are more aspects of the life of a creature that deserve moral attention besides its suffering. Among such aspects we can mention the creature's well-being, its autonomy and its right to self-fulfilment. However, there is a crucial difference between the significance such aspects take in the positions of the two camps. For the impartialist, the status of such interests *as* interests of a creature must be determined independently of the contingent concerns and sensitivities we have in regard to the creature. For Singer, for example, 'Our distress [about the killing of some creature] is a *side effect* of the killing, not something that makes it wrong in itself.'[5]

Singer insists that the concerns we happen to have are simply irrelevant for moral reasoning and that valid moral reasoning must limit itself to identifying others' interests. While Singer is struggling to allow that, if put to the test, he would save his daughter Naomi, and not his dog Max, *because* she is his 'lovely baby daughter',[6] he insists that this 'because' lacks any moral value, and any valid form of justification of his choice must be given in terms of facts about her capacities and potentialities not shared by the dog. Thus, any appeal to concerns about other's interests and well-being, for Singer, is relevant to the moral question only in so far as it helps to *identify* interests that are out there, independently of such concerns. Singer cannot recognize anything but bare emotions, devoid of reason, in a position that allows the concerns we have about others to enter the description of the moral facts.[7] For moral reasoning to be a valid case of reasoning, Singer insists, the moral significance of the circumstances in question must be cashed out independently of any contingencies about the concerns we happen to have. We can use Wittgenstein's distinction between *symptoms* and *criteria* to clarify the debate: Singer insists that our moral intuitions (of course, Singer might choose not to call them 'moral') – that is, our concerns about others' interests – are *symptoms of*, not *criteria for*, the moral value of those others and of our moral obligations towards them.[8] The Wittgensteinian, by contrast, takes such concerns to be criteria for valid moral principles.

3. The Preferential View

We can call this Wittgensteinian conception the 'Preferential View'. According to this view, the starting point of moral reasoning is in respecting our preferences towards

[4] See, for example, Yuval Eylon, 'Virtue and Continence', *Ethical Theory and Moral Practice*, 12 (2009), 137–51.

[5] Peter Singer, 'Reflections', in J. M. Coetzee, *The Lives of Animals* (Princeton, NJ: Princeton University Press, 1999), 89.

[6] Ibid., 87.

[7] Cf. Diamond, *The Realistic Spirit*, ch. 11, where Diamond extracts the image of the nature of moral thinking underlying such a conception; see, for example, 305. Diamond addresses Singer's own conception in various essays, notably 'The Difficulty of Reality and the Difficulty of Philosophy', in *Philosophy and Animal Life* (New York: Columbia University Press, 2008), 50. See also Singer, 'Reflections', 91.

[8] For Wittgenstein's distinction between criteria and symptoms, see his *Philosophical Investigations*, 2nd edn (Oxford: Blackwell, 1958), §354. Cf. Singer, 'Reflections', 89.

some over others, rather than by attempting to justify such preferences on the basis of some, supposedly, firmer grounds. What could be said in favour of such a view?

The impartialist's goal is highly ambitious, namely: to establish foundations for morality on firm, objective grounds. However, there are strong reasons to believe that this goal cannot be met, and moreover, that an attempt to meet it is doomed to result in a moral disaster. I will argue that the impartialist's craving for foundations, objectivity and neutrality could be fulfilled only at an intolerable cost – namely, the cost of our being burdened with a conception that is hardly recognizable as a conception of morality at all. To see this more clearly, we will first have to formulate in a more precise manner the challenges the impartialist presents to the preferential view.

Cora Diamond describes the impartialist (though Diamond does not use this particular term) as committed to the following picture: Human beings have natural inclinations, which are not based on reasons (they prefer cats over cows, members of their own ethnic group over members of another, etc.). Such inclinations, the impartialist holds, are no more than mere manifestations of subjective sentiments unless they are grounded in facts and general principles, which are independent of their frames of mind (or better, their frames of heart).[9]

We can make the impartialist's picture more explicit: The impartialist holds that (morally) caring for cats while unhesitatingly consuming cows is justified *only* if the following conditions are met:

Let M be the moral attitude in question, supposedly applying to cats (or dogs) but not to cows.

(i)   There is some feature F shared by cats, but not by cows.
(ii)   There is some moral principle according to which holding the moral attitude M towards S is justified if and only if F is a feature of S.[10]

However, the impartialist argues that there is no such feature F. Therefore, holding the attitude M towards cats but not towards cows is unjustified. Therefore the preferential view is false.

There are two distinct threats the impartialist takes the preferentialist to be vulnerable to: First, there is the threat of sentimentalism (or anthropomorphism), and hence the objection might be raised that our preferences are not sufficient for the detection of certain justified moral claims of others. It might be noted, for example, that

---

[9]   Diamond, *The Realistic Spirit*, 293.
[10] We can add the following thesis to the impartialist's argument: (iii) It is unlikely that such a feature F distinguishes cats from cows. According to an influential line of argument, the reason it is unlikely is that other cultures do not share the moral attitudes we have towards cats, while they have *these very attitudes* towards cows, and vice versa. Limitations of space prevent me from dealing with this argument explicitly, but I hope the argument of the present essay suffices to show why this common claim about 'other cultures' is misguided. In any case, I believe the argument is based on a false description of the cultural differences concerning attitudes to animals. It is assumed, for example, that the special status cows have in Hinduism parallels the role pets have in many urban societies. But a better comparison of the status of cows in Hinduism is to the status of holy objects in some religions, not to that of pets.

the mere fact that *we* happened to favour some creature over others lacks any moral significance. Children, for example, often express deep concerns about the condition of their teddy bears exceeding their concerns about the well-being of their fellow friends. Such concerns have nothing to do in morally *justifying* the favouring of teddy bears over the human friends.

Secondly, there is the threat of a certain kind of moral blindness, since our preferences do not necessarily detect the moral claims of others. Possessing a given set of preferences does not necessarily equip their possessor with adequate capacities for the recognition of justified moral claims of others. Thus, preferring some to others could easily blind one to at least some of the moral claims others make upon one.

To secure ourselves from such threats, the impartialist insists that morality must be given firm grounds, independent of the tendencies of our hearts, which underlie our preferences.

4. Impartialist worries

The difficult thing here is not, to dig down to the ground; no, it is to recognize the ground that lies before us as the ground.[11]

The first worry of the impartialist is that in relying on her own preferences an agent could lose track of morality altogether. Preferences as such, she argues, are arbitrary. We might prefer teddy bears to people, enjoy torturing the innocent, etc. In other words, preferences are simply the wrong type of things to base morality on. They do not belong to the *order of justification* – they are not the types of things that have the power to justify or to be justified.[12] The fact that we often like some people or creatures more than others is nothing but a fact about our subjective life.

This picture of preferences and other subjective states as arbitrary and lying beyond justification, however, should be discarded. One of the most influential arguments against this picture is found in Wittgenstein's passages often referred to as 'The Private Language Argument'.[13] The upshot of Wittgenstein's argument is that the idea of the independence of a subjective realm or order of justification is incoherent. We can illustrate that by the help of a comparison which Wittgenstein makes. In *Philosophical Investigations*, §293, Wittgenstein asks the reader to imagine some community in which everyone has a box in which they keep a beetle. Now, no one is allowed to look in anyone else's box, only in his or her own. Over time, people talk about what is in their boxes and the word 'beetle' becomes a common word used in their shared

---

[11] Ludwig Wittgenstein, *Remarks on the Foundations of Mathematics*, 3rd edn (Oxford: Blackwell, 1978), VI, §31.

[12] See, for example, McDowell, *Mind and World*, 29, 133–5; *The Engaged Intellect* (Cambridge, MA: Harvard University Press, 2009), 249, 253. McDowell uses the image of 'the order of justification' in the context of sensory experience, but it is equally applicable to the case of 'inner experience'.

[13] See Wittgenstein, *Philosophical Investigations*, §§243–315.

language. However, Wittgenstein argues that we are only under the illusion that we have successfully described a possibility here.

Wittgenstein argues that the thing in the box is neither necessary nor sufficient for the use of 'beetle' in the public language. It is not sufficient because whatever is in fact in the box – whether it is a particular object, or something that is constantly changing, or whether it is empty – the speaker's use of the term 'beetle' would refer to *that thing* (or those things, or to nothing).[14] So no content would have been given to the claim that the term refers to what is in the box. A term that can refer indiscriminately to anything is not a genuine referential term. And if the term is not referential, the thing it supposedly refers to simply drops out: 'No, one can 'divide through' by the thing in the box; it cancels out, whatever it is.'[15] On the other hand, whatever is supposedly in the box is not necessary for the way the term is used in public language because there is no way in which the thing in the box could play any role in the considerations governing a proper application of the term in public use. The content of the box could not contribute to the battle of considerations used in justifying or in criticizing a given application of the term. 'The thing in the box has no place in the language-game at all; not even as a *something*.'[16]

Now, what is true about 'beetle' in Wittgenstein's imagined community is surely true of the use of a psychological term, such as 'desire', 'preference', etc., when understood in abstraction from 'the order of justification' – that is, the different considerations, reasons, grounds and doubts users appeal to in justifying or criticizing the use of such terms in public language. But such an abstraction underlies the accusation of the impartialist. The impartialist argues that the preferentialist is committed to sentimentalism because she bases her account on mere preferences, which the impartialist holds to be purely subjective states, incapable of playing any justificatory role. But the Wittgensteinian argument shows that such a picture of preferences and other subjective contents is an illusion. It is not a picture of anything. The only reason we confusedly assume it to be a picture of content is because we are doomed to describe it with the help of terms from our shared, public language. But when this putative picture is seen to be illusory, the constituents of the mental need no longer be regarded as alien to the order of justification, and hence the basis of the impartialist's objection diminishes.

This reply to the impartialist's accusation, given here only in brief outline, should be taken with some caution. The claim that our preferences and other mental states are not alien to the order of justification should not be understood as implying that we are never prone to different errors as a result of relying on our preferences, sentiments, etc. The point is, however, that even occasions in which we fall victim to such biases do not sharply remove the attitudes we rely on from the order of justification. In referring to such cases as 'failures' it is implied that, when properly exercised, our preferences have the power to justify and to be justified. A failure or error of preference is nothing less than a failed exercise of a capacity, which is primarily a capacity to be sensitive

---

[14]   See ibid., §293.
[15]   Ibid.
[16]   Ibid.

to different patterns of justification, including ones involving common standards of moral justification.

The second of the impartialist's concerns, which I noted briefly above, is the worry that not only are our preferences not *sufficient* for morality, they are not even *necessary* for it. Resting our moral outlook on our preferences makes our moral thinking prone to the threat of ignoring some valid moral claims of others simply because they have the misfortune of not being among the creatures (human or otherwise) we happen to like. In the absence of firm objective grounds underlying our moral outlook, there is no warrant that our preferences equip us with proper sensitivities enabling us to discern what morally matters.

However, the impartialist insistence on the need for firm, 'heart-indifferent' grounds could not be met. The problem is that the idea of grounding by independent principles of morality is hardly recognizable as a case of grounding at all, even less so as a case of 'moral' grounding. The impartialist's accusation, that the cat/cow distinction (or the dog/cow distinction) falls short of supplying the necessary 'firm grounds', is based, I will argue, on an unattainable myth. To see that, let us consider a putative moral principle of the following form:

MP: We ought to minimize suffering in the world.

The principle MP supposedly grounds our attitudes towards any creature, independently of whether we care about their suffering. But it is not clear that MP could serve as a ground for any moral outlook whatsoever. The problem is that putting suffering *simpliciter* as the principle of action is forgetting why we *care* about suffering – what motivates us to aim at preventing it. As a general rule for action MP does not have, in itself, any justificatory force. Of course, the impartialist's choice to place suffering at the heart of a purported moral principle is not random – suffering is something we often *care* about, and other things being equal most of us would wish to minimize. By using the term 'suffering' we already connect the person or the creature we describe to 'what human life is' – that is, to the complex network of concerns and engagements involved in interacting with our fellow human beings.[17]

In describing a creature as suffering we are already placing her in a rich network of attitudes, such as wishes, hopes, beliefs, as well as related concepts such as those of pain, compassion, cruelty, etc. The concept of 'suffering' is deeply rooted in the life of a person, or of a creature, and the propriety of applying it depends on many aspects of the subject's life. The search for firm grounds, in the form of a universal principle detaches 'suffering' from such a complex network, and thus detaches it from *why we care* about suffering. The resulting purported moral principle is suggestive only because it borrows its appeal from the force the concept of 'suffering' has when entrenched within the complex idea of the life of a creature. By *endorsing* the principle MP when considered

---

[17] Cf. Diamond, *The Realistic Spirit*, 325–6: 'The ways in which we mark what human life is belong to the source of moral life, and no appeal to the prevention of suffering which is blind to this can in the end be anything but self-destructive.'

in abstraction from the complex role 'suffering' plays in forming our judgements about and interactions with others, the principle turns into an arbitrary *cause* for action and belief-formation rather than a moral justification for action, and there is no reason to expect that the two would coincide.[18]

In fact, it is not even clear that what we call 'suffering' in our ordinary interaction with others is discernible from the remote point of view the impartialist recommends. The question whether particular events in the lives of a person involve suffering depends on more than the amount of pain the creature undergoes (surely, on more than the number and type of neurons triggered in her brain).

5. Attitudes versus opinions

What account can be given of the moral status of animals, once the search for a neutral perspective and for firm grounds is given up? We can recall, in this context, Wittgenstein's advice: 'don't think, but look!'[19] In the context of the present discussion, the advice should be to observe carefully the characteristics of our interactions with animals, at the different types of considerations we appeal to in debating animals' moral status and at the forms of concern we direct to animals, etc.

There are many similarities, as well as striking differences, between our attention to animals and our attention to fellow human beings. (Similarly, there are many similarities and differences between our attention to some human beings versus our attention to others.) Tracing such connections allows placing morality in a general humanistic framework to which it naturally belongs, even when the subjects in question are non-human. Thus, the question of the moral status of animals is revealed to be an extension of the question of the moral status of human beings.

The attitudes involved in the ways many of us interact with and think about our dogs and cats, for example, have a deep person-like character:[20] a dog is often described as a loyal friend and neglecting it amounts to a betrayal; every cat has a unique character, much like humans; a dog can be happy or sad, depending on the behaviour and mood of its owner, and a cat can be jealous, much like a human child, when her owner nurses a new baby; cats and dogs are the close companions of many, and the only companions for some. Such characteristic forms of expression seem to stand at the heart of our concern and care (or at least, the concern and care some of us have) towards dogs and cats.

In Part II of the *Philosophical Investigations*, Wittgenstein writes: 'My attitude towards him is an attitude towards a soul. I am not of the *opinion* that he has a soul.'[21]

---

[18] The impartialist, of course, is unlikely to find this objection threatening. After all, she denies the requirement that moral demands should coincide with our intuitions. When put like that, it is clear that there are here two radically conflicting conceptions of morality, a choice between which cannot be made on the basis of arguments. (I would like to thank Nora Hämäläinen for raising this point.)

[19] Wittgenstein, *Philosophical Investigations*, §66.

[20] Cf. Diamond, *The Realistic Spirit*, 324.

[21] Wittgenstein, *Philosophical Investigations*, Part II, iv, p. 178.

The idea of 'attitudes' versus 'opinions' can be illustrated by such characteristic forms of attention to animals. We can say that Wittgenstein's proposal, when applied to the case of animals, is to look at our forms of relations to cats and dogs not as resulting from opinions or as based on hypotheses that need to be supported by evidence. Taking my dog to be a loyal companion is not an opinion I have about my dog. An opinion presupposes the truth of some hypothesis about the type of creature my dog is; but my *attitude* towards my dog cannot be refuted as though it were a hypothesis.

It is not that my attitude to my dog cannot be challenged, but the sorts of considerations that would challenge taking my dog to be a loyal companion are of a piece with the typical considerations involved in doubting that someone is a loyal companion.[22] Such considerations might include, for example, evidence such as: 'she betrayed me', or 'she doesn't really care about me at all', as when I learn that my dog is indifferent to my absence or presence as long as there is someone around her. Of course, not *all* such forms of expression are applicable to dogs, but that does not mean that the grounds that could make me stop taking my dog to be a loyal companion are of a substantially different form from the grounds that typically make us stop taking a human friend to be loyal.

Yet there are also considerations that could influence us in the case of dogs that are more peculiar to the case and that would not apply to human beings. For example, a loner, who never succeeded in finding trust in people might find a true human friend, one day, and feel it is only now he understands what 'true friend' really means. He might realize that while his dog was a great comfort to his loneliness, it was nothing like a true friend. But even this case could hardly be described as a *discovery* of some fact underlying a true friendship that is absent in the case of dogs. In fact, we can similarly imagine someone who by finding a true friend and learning of the true virtues of friendship, realizes that he never had a true human friend before, though he always took himself to have many.[23]

6. Pets and persons

When we say we don't eat pets, for example (or that we don't kill prisoners), we imply something about who *we* are – rather than about what is *right* for any being in the cosmos to do under any possible circumstances, as a result of some firm moral principle. It is not that we believe that for others it is *right* to kill prisoners or to eat pets. Rather, referring to some people as 'prisoners' and to some animals as 'pets' is already making the first step in placing them within a complex network

---

[22] In that sense, we should be cautious of reading Wittgenstein's distinction between opinions and attitudes as an attempt to demarcate a sharp dichotomy. It should not be concluded from stating that *I am not of the opinion that you have a soul* that the attitudes I take towards you as a human being are insensitive to the influence of a variety of beliefs, opinions and hypotheses, and that they could not be revised in different ways.

[23] Niklas Forsberg suggested that we can also imagine the reversed case, of a man who came to think that he has never had a real friend in his whole life, until he got a dog.

of relations to various beliefs, values, practices, etc., relations that already carry various implications about how we think they should be treated. Calling an animal a 'pet' is already making some person-like comparisons, and placing it in a position that excludes it from the food chain, making the idea of eating it rather grotesque.[24]

The way to put it is not that, first, we recognize some independent *similarity* – some common features shared by humans and pets, describable from the perspective of the 'cosmic exile'[25] and in light of which the 'person' vocabulary is applicable to pets – and *then* we are justified in treating pets in a manner similar to that in which we treat people.[26] Diamond illustrates this point by saying:

> A pet is not something to eat, it is given a name, is let into our houses and may be spoken to in ways in which we do not normally speak to cows or squirrels. That is to say, it is given some part of the character of a person.[27]

It is crucial that it is *we* who *give* pets this person-like status. However, giving pets such a status is not something we choose to do, something we could willingly have chosen not to do. Rather, it is a result of the fact that we *grasp the behaviour of dogs and cats and our interactions with them under a person-like mode of presentation*. Hence, there is no need to demand a general justification for ascribing a human-like character to some pets. It is, rather, that some animals take particular roles in our social and private lives, and such roles make the human analogies apt. Denying that there is a need for justification does not mean that there is nothing 'right' about taking some animals to have a human-like character. Recall Wittgenstein's words in *Philosophical Investigations*, §289: 'To use a word without a justification does not mean to use it without right.' In other words, applying the human vocabulary to pets is not based on *another* justification besides the fact that such a vocabulary comes naturally to us, given the role such animals play in our lives and the forms of interaction that we have with them.

[24]   Possibly, there is room to make some distinctions within the general category of 'a pet'. The case of some pets, and in particular dogs and cats, is peculiar because dogs and cats are the types of creatures that often can play this human-like role. Rabbits, hamsters and iguanas can play it to a lesser extent. But it must be emphasized that the argument of the present essay also shows that there is no non-question begging answer to the question what must be true of a creature, so that it could be subject to this more complex 'human-like' attention and participate in such unique forms of interaction.

[25]   I borrow this expression from John McDowell, 'Anti-Realism and the Epistemology of Understanding' [1981], reprinted in his *Mind, Knowledge, and Reality* (Cambridge, MA: Harvard University Press, 1998), 330. McDowell himself borrows it from W. V. O. Quine, *Word and Object* (Cambridge, MA: MIT Press, 1960), 275.

[26]   Compare the thesis that there is no fact of the matter that justifies saying that a person's behaviour (described from the 'cosmic exile') is a case of adding rather than of quadding. (This formulation of Wittgenstein's problem is influenced by 'Wittgenstein on Following a Rule,' in John McDowell, *Mind, Value and Reality* (Cambridge, MA: Harvard University Press, 1998), 221. Of course, 'quadding' is defined by Saul Kripke in *Wittgenstein on Rules and Private Language* (Cambridge, MA: Harvard University Press, 1982), 9.)

[27]   Diamond, *The Realistic Spirit*, 324.

We think about and interact with some animals in ways that share many aspects of the ways we think of and interact with our fellow human beings. Maybe the most important manifestation of such a human-like grasp of some non-human creatures is that it enables us to empathize with a creature – to see many aspects of its life, its suffering, its interactions with other creatures, etc., as meaningful events in a life; though much simpler than a human life, still sharing enough with it so that we can understand it and relate to it.[28] This ability to relate to the other's life is essential to our grasp of other human beings, though such ability could degenerate, as with a racist that fails to see some other as a fellow human being. But that is precisely the sense in which it makes sense to say that, at least in some cases, a racist approach to others is misguided: it consists of *a failure* – the failure to grasp others in the same rich spectrum of sensitivities in which we grasp other human beings. In this case, as in the case of animals, talking about 'failure' versus 'success' does not mean that there is a justification that the racist failed to recognize. The racist is not making a *factual* mistake about some common feature shared by those she is racist towards and those to whom she is not. *Given* our approach to the family of human beings, that includes all races and genders, and to a lesser extent some animals, we can say that the racist approach involves a moral failure – the failure to exercise some of the sensitivities that are characteristic of a moral approach to others.

Grasping a situation involving some creature under person-like concepts is grasping it as holding various relations to reasons, motivations, beliefs, etc. If we *see* a fly trying to fly out of the window – that is, if we already grasp the situation *that way* – then, by this token, we can understand a motivation someone might have to open the window and let it out. It is not that we must share this motivation, but rather, grasping this motivation of someone enables us to recognize a reason for letting the fly out, even if we are not motivated by this reason. In grasping it, suddenly, we can understand what it means to have such a motivation; the statement ascribing the motivation to the individual concerned will not strike us as a sheer piece of nonsense. Recognizing and even sharing the motivation to let the fly out does not mean that we made a first step in formulating a new general rule for action, such as 'Let out anybody trying to leave the room'; it is rather that there is now a *comparison* we find apt. By accepting the comparison, we are committing ourselves (or better, we find ourselves committed) to various links, connections, motivations, etc. Thus, if someone is in sadistic mood

---

[28] In his recent monograph, *Love and Death* (Milan: Silvana Editoriale, 2010), the Japanese photographer Nobuyoshi Araki documents the last days in the life of his beloved cat Chiro. In this set of images Araki is nearly repeating the compositions of the images in his *Sentimental Journey, Winter Journey* (Tokyo: Shinchosha, 1991), a work documenting the last days in the life of his wife, 20 years earlier (Chiro was already a central figure in the early work). Fascinatingly, the images of the cat's suffering revive the suffering of Yoko, Araki's wife, in the earlier work. The young Yoko's sadness and pain, and the difficulty of the spectator to find in it anything but a remote loneliness, suddenly gains new heartbreaking force in the form of the explicit facial expressions of the helpless cat, pleading for some comfort from the photographer – his loving companion – Araki. It is possible that the view of some observers (myself included) of the work is limited as they simply miss some of the facial nuances and conventions of expressions of emotions manifested by Yoko. This, I believe, is an interesting case, in which the image of the cat enables at least some viewers to see the human in an even more human-like manner than she otherwise could have done.

she might open the window and close it just before the fly escapes; she might find it funny, while there would be nothing funny (or if there were, it would be funny in a very different way) if it had been a leaf blowing in the wind and hitting the window. Closing the window just before a leaf hits it would have been a very different game.

7. Back to the everyday

The aim of this essay has not been to convince anybody that cats and dogs can be one's friends, companions or sometimes part of one's family. Neither has it been to find any fault in someone's decision to let her cow sleep on her lap while watching TV. Asking for a justification for not eating cats is wrongheaded. A cat is a pet and we don't eat pets, and we certainly don't eat a friend, not even someone else's friend. This is not a consequence of a general independent moral principle. Someone who understands the comparison between cats and friends but demands additional moral grounds for not eating cats is probably best characterized, not as a sceptic, but as morally evil or, at least, heartless.

'What *we* do' (in philosophy), says Wittgenstein, 'is to bring words back from their metaphysical to their everyday use'.[29] We can think of the use of 'morality' in the sense of the search for firm grounds, from a heart-indifferent point of view, as a case of what Wittgenstein refers to as a 'metaphysical use' of an expression. The problem with a metaphysical use of an expression is not that in using expressions in such a way we express a special type of meaning, a *metaphysical* meaning. Rather, the problem is that such uses commonly involve the pretence of reaching a deeper, underlying sense of the expression when, in fact, all that is achieved is the use of the expression in a more precise, but at the same time also a more limited and technical, way – ignoring many aspects of its common use.

So, we should take the advice 'to bring a word back to its everyday use' as the advice to remind ourselves of other legitimate common uses of an expression that are left out by the pretence to give the expression a well-defined use. We can think of the impartialist as such a metaphysician – one who insists on reserving the term 'morality' for the context of impartiality. The argument above shows the limitation of such a proposal. It shows that limiting 'morality' to such a context renders it unrecognizable as anything worthy of being called 'morality'.

What type of uses are the sources of the temptation to identify morality with a crippled, metaphysical use? I believe the answer should be found in the idea that morality should save us from the weaknesses of our hearts. There are, in our lives, many occasions on which we are blind, biased, distorted, perverted, racist, chauvinist, etc. And these occasions, when our hearts fail to see the other as a human, are those in relation to which it is tempting to conceive of morality as a realm of the heart-indifferent – a detached and general outlook. It is these occasions which make us

---

[29]   Wittgenstein, *Philosophical Investigations*, §116.

believe that any use of our heart in moral reasoning is a moral failure. This temptation to conflate morality with impartiality is ironically mocked in Oscar Wilde's *An Ideal Husband*, where Mrs Cheveley declares: 'Morality is simply the attitude we adopt towards people whom we personally dislike.'[30] The purpose of the present essay has not been to vindicate Mrs Cheveley's exaggerated statement. Rather, it has been to show that the extreme position expressed in that statement can be avoided without succumbing to impartialism; that is, without underestimating the vital role that preferences play in our moral lives, with people and with animals.[31]

---

[30] Oscar Wilde, *An Ideal Husband* [1895]. See *Collected Works of Oscar Wilde* (Ware: Wordsworth Editions, 1997), 630.

[31] I would like to thank Yuval Eylon, Dalia Dray, Alice Crary and the editors of this volume for their helpful comments which improved this essay greatly. I also want to thank the Philosophy Department at the University of Haifa and Haifa Center for German and European Studies for their generous support that enabled me to dedicate my time to the research.

10

Second Nature and Animal Life

Stefano Di Brisco

It is not uncommon among philosophers to say that humans and animals have different kinds of mindedness. What they mean is that, even if we are right in acknowledging animals as self-moving and responsive to features of their environment, only human beings can properly be said to be 'rational beings'. As we may put it, we share with animals a perceptual sensitivity to the environment, and differ from them in our rationality. Thus, for example, both humans and animals can feel a pain, but only a human being can tell others how that pain occurred. It is very difficult to spell out the peculiarity of the human condition, and even more to assess its relevance to our ethical commitments towards animals. So the disagreement among philosophers on this matter is often about how to understand and assess the idea of human uniqueness in nature.

I'll be concerned in this essay with John McDowell's original account of this topic, as it figures in his book *Mind and World*.[1] McDowell offers a conception of human mindedness that allows us to make sense of the question about the human uniqueness in nature outside the Cartesian framework in which it is often asked. I will focus on how McDowell arrives at facing the human–animal distinction and the role it plays in the structure of his book: this will give me the occasion to develop some critical evaluations of McDowell's method concerning (what I take to be) a tension between his philosophical quietism[2] and the kind of generality he is aiming at. My discussion is also intended as an illustration of the importance of the starting point for philosophical reflection on humans and animals.

1. McDowell's *Mind and World*

I will begin by briefly sketching the overall topic of *Mind and World*. In the lectures that constitute this book, McDowell aims at uprooting a persistent dualistic picture

---

[1] John McDowell, *Mind and World* (Cambridge, MA: Harvard University Press, 1996).

[2] Since the term 'quietism' can be used in many ways, I want to specify my use of it here. In my usage it indicates a particular mode of philosophical activity McDowell inherits from Wittgenstein. At §127 of his *Philosophical Investigations*, 3rd edn (Oxford: Blackwell, 2001), Wittgenstein says that 'The

of the relation between mind and world that gives rise to (what he considers) familiar anxieties about the nature of that relation and of its components.[3] In general, this picture is historically identified with the post-Cartesian tradition, according to which thought is separated from reality, reason from nature and norms from facts. This picture represents our mental activities (such as understanding, knowing, thinking, etc.) as a bounded space, which occasionally can make contact with the external

work of the philosopher consists in assembling reminders for a particular purpose,' as opposed to putting forward distinctively philosophical theses (see *Philosophical Investigations*, §128). 'Assembling reminders' means saying to someone something she already knows, but perhaps has forgotten. The kind of quietism McDowell associates with Wittgenstein's later work is thus contained in the idea that 'philosophy leaves everything as it is' (*Philosophical Investigations*, §124), where this remark should not be taken as a defeat in the face of genuine philosophical problems (compare §118). Rather, the avoidance of substantive philosophy is the result of a diagnostic activity that recognizes many philosophical questions (such as, e.g. 'How is meaning possible?') as illusory appearances of problems. Following Wittgenstein's (and McDowell's) medical metaphor, many traditional philosophical questions are pathologies of the intellect in need of 'therapy' (compare *Philosophical Investigations*, §133 and §255). The therapeutic purpose for which the philosopher assembles reminders is precisely that of relieving us of the pressure of bad philosophical pictures that make certain questions urgent for us. In a recent paper McDowell writes: 'Wittgensteinian quietism involves being suspicious of philosophers' questions, before we even start interesting ourselves in the specifics of how they are answered. If someone invites us into substantive philosophy by, say, asking something of the form "How is such and such possible?", we should not at once embark on trying to give a positive philosophical account of such and such, whatever it is. First we should ask why we are expected to find a difficulty in the possibility of such and such, whatever it is. Often the best answer that can be given will seem to carry conviction only to the extent that it induces us to forget something obvious. Revealing such a defect in the supposed pretext for the "How possible" question, and so entitling ourselves not to have to bother with it, at any rate if that is the ground on which we are invited to find it pressing, is a distinctive kind of philosophical achievement. . . . Quietism does indeed urge us not to engage in certain supposed tasks, but precisely because it requires us to work at showing that they are not necessary. And it is indeed work. Therapeutic philosophy is designed to spare us the travails of positive philosophy, but it has its own difficulties. There is no guarantee that it will be easy to uncover a forgetfulness of something obvious, underlying the conviction of being under an intellectual obligation to engage in one of those tasks. . . . So this kind of philosophy needs a precise and sympathetic appreciation of the temptations it aims to deconstruct. There is no question of quickly dismissing a range of philosophical activity from the outside' (John McDowell, 'Wittgensteinian "Quietism"', *Common Knowledge*, 15 (2009), 365–72, at 371–2). McDowell has previously discussed Wittgenstein's 'quietism' in his 'Postscript to Lecture V', in *Mind and World*, 175–80. See also John McDowell, 'How Not to Read *Philosophical Investigations*: Brandom's Wittgenstein,' in *The Engaged Intellect* (Cambridge, MA: Harvard University Press, 2009), 96–111. For a general characterization of Wittgenstein's philosophy as 'therapeutic', see Alice Crary and Rupert Read (eds), *The New Wittgenstein* (London: Routledge, 2000), especially the 'Introduction' by Alice Crary, at 1–18.

[3] McDowell's talk of 'anxieties' in this context expresses his commitment to Wittgenstein's idea that often, in philosophy, we are confronted with illusory appearances of problems (see note 2 above). In the 'Introduction' McDowell writes: 'My aim is to propose an account, in a diagnostic spirit, of some characteristic anxieties of modern philosophy - anxieties that centre, as my title indicates, on the relation between mind and world. Continuing with the medical metaphor, we might say that a satisfactory diagnosis ought to point towards a cure. I aim at explaining how it comes about that we seem to be confronted with philosophical obligations of a familiar sort, and I want the explanation to enable us to unmask that appearance as illusion. / It matters that the illusion is capable of gripping us. I want to be able to acknowledge the power of the illusion's sources, so that we find ourselves able to respect the conviction that the obligations are genuine, even while we see how we can, for our own part, reject the appearance that we face a pressing intellectual task' (*Mind and World*, xi). For a clear description of a Wittgensteinian style of criticism that points to bad philosophical pictures as sources of illusions, see James Conant's 'Introduction' in Hilary Putnam, *Words and Life* (Cambridge, MA: Harvard University Press, 1994), in particular xlvi–lviii.

world. It happens that thought, when it is thought about something, touches reality. According to this picture, the objectivity of our (empirical) thought depends on the possibility of finding a link between our concepts and the external reality. Such a link is required if we want to allow ourselves the capacity of making warranted judgements about the world. It is natural to find such a link in our sensibility: in standard conditions, our senses bring us into contact with the world. Consider for instance the case of visual experience: if one is not subject to hallucinations, nor under optical illusions, one *sees that* the table is, say, white. 'In experience', McDowell writes, 'one finds oneself saddled with content', in a way that leaves one with no 'choice in the matter'.[4] And we are justified in judging that the table is white if the table is indeed white. It is thus a requirement for there to be warranted judgements about the world that experience, functioning as a 'tribunal',[5] mediates the way our thinking bears on external reality, so that it is *correct* or *incorrect* to judge, for instance, that the table is white according to whether or not it is indeed white. McDowell explains this point as follows:

> A belief or judgement to the effect that things are thus and so – a belief or judgement whose content (as we say) is that things are thus and so – must be a posture or stance that is correctly or incorrectly adopted according to whether or not things are indeed thus and so. . . . This relation between mind and world is normative, then, in this sense: thinking that aims at judgement, or at the fixation of belief, is answerable to the world – to how things are – for whether or not it is correctly executed. . . . And now, how can we understand the idea that our thinking is answerable to the empirical world, if not by way of the idea that our thinking is answerable to experience? How could a verdict from the empirical world – to which empirical thinking must be answerable if it is to be thinking at all – be delivered, if not by way of a verdict from (as W. V. Quine puts it) "the tribunal of experience"?[6]

We begin to understand how for McDowell the dualistic picture gives rise to an anxiety about the possibility for there to be thinking 'answerable to how things are', if we consider a contrast between two rather intuitive *desiderata* about the relation between mind and world: on the one hand, we conceive our minds as free, that is as being able to be governed by rational norms and relations, and not merely by the binding force of physical (causal) laws. This is the reason why McDowell adopts the Kantian term 'spontaneity' to characterize our conceptual faculty.[7] Our conceptual capacities belong

---

[4]   McDowell, *Mind and World*, 10.
[5]   For the metaphor of the 'tribunal of experience' see W. V. O. Quine, 'Two Dogmas of Empiricism', in *From a Logical Point of View* (Cambridge, MA: Harvard University Press, 1961), 20–46, at 41.
[6]   McDowell, *Mind and World*, xi–xii.
[7]   McDowell writes: 'When Kant describes the understanding as a faculty of spontaneity, that reflects his view of the relation between reason and freedom: rational necessitation is not just compatible with freedom but constitutive of it. In a slogan, the space of reasons is the realm of freedom' (ibid., 5).

to what Wilfrid Sellars called 'the space of reasons',[8] that is, they operate in a normative context that defines the rational structure of our judgements and beliefs. The space of reasons must be distinguished from the 'realm of law', which is the sphere of activity of empirical sciences. This 'dichotomy of logical spaces'[9] is the acknowledgement of a difference in kind between concepts we employ when we locate something in the normative context of the space of reasons, and concepts we use in an empirical description when we situate something in the realm of law.

On the other hand, we want our thoughts to be guided by the reality outside our minds, so that the spontaneity of our conceptual faculty does not result in a frictionless absolute. For if 'thinking . . . is not constrained from outside the conceptual sphere, that can seem to threaten the very possibility that judgements of experience might be grounded in a way that relates them to a reality external to thought'.[10] If our freedom were unconstrained by the world, we could not make sense of ourselves as knowing or representing the world. If the world cannot make contact with our concepts, then our concepts cannot be about the world. The problem, then, is finding an acceptable way for the world to constrain the freedom of our conceptual faculties.

As I have noted, it is natural to think that our exercises of concepts are constrained by sensory impressions the world makes on us. These sensory impressions are taken to be what make up experience. But, following Quine's metaphor, we have also introduced the requirement that experience functions as a tribunal, to which our thought must be answerable. In order for experience to function as a tribunal, experience must play a normative role in our thinking: it must provide *rational* constraints on our exercises of concepts. But can we really understand the relation between thought and impressions in terms of 'answerability to a tribunal'? Sensory impressions are natural phenomena; as such, they belong to natural law. The concepts of 'thought' and 'answerability to experience' are instead normative ones, and thus operate in the space of reasons. So how can experience, conceived as made up of sensory impressions, function as a tribunal that delivers its verdicts on our judgements and beliefs? In other words, how can mere sensory impressions constitute a *justification* for our empirical thought? As McDowell observes, 'the logical space in which talk of impressions belongs is not one in which things are connected by relations such as one thing's being warranted or correct in the light of another. So if we conceive experience as made up of impressions . . . it cannot serve as a tribunal, something to which empirical thinking is answerable'.[11]

To sum up what has been said so far: in the context of a reflection about the world-directedness of thought, we find ourselves subject to the pressure of a pair of thoughts that cannot be renounced: on the one hand, we want our thought to be constrained by the world by being answerable to the 'tribunal of experience'; this is required if we want to make sense of experience as providing a justification for our empirical judgements

8 Wilfrid Sellars, 'Empiricism and the Philosophy of Mind,' in *Minnesota Studies in the Philosophy of Science*, Vol. 1, ed. Herbert Feigl and Michael Scriven (Minneapolis: University of Minnesota Press, 1956), 253–329.
9 McDowell, *Mind and World*, xv and *passim*.
10 Ibid., 5.
11 Ibid., xv.

and beliefs. On the other hand, there seems to be an insurmountable difficulty in conceiving experience as a tribunal: how can mere sensory impressions constitute rational constraints on our free exercises of concepts? According to McDowell, the pressure of this pair of thoughts generates an anxiety about the possibility for our mind to make contact with reality. He writes:

> Fully developed, of course, such a combination [of thoughts] would amount to an antinomy: experience both must . . . and cannot . . . stand in judgement over our attempts to make up our minds about how things are. . . . With an inexplicit awareness of the tension between such a pair of tendencies in one's thinking, one could easily fall into an anxiety of a familiar philosophical sort, about that directedness of mind to world that it seemed we would have to be able to gloss in terms of answerability to how things are. In such a position, one would find oneself asking: "How is it possible for there to be thinking directed at how things are?" This would be a "How possible?" question of a familiar philosophical kind; it acquires its characteristic philosophical bite by being asked against the background of materials for a line of thought that, if made explicit, would purport to reveal that the question's topic is actually not possible at all.[12]

In McDowell's exposition, the anxiety takes the form of an oscillation between two exclusive and opposed philosophical positions, the Myth of the Given and Coherentism. 'Myth of the Given' is the label Wilfrid Sellars coined to indicate (and criticize) the idea that the ultimate grounds of our empirical judgements are impressions the world makes on us, as possessors of sensibility.[13] The Given is supposed to be something non-inferentially arrived at, an unmediated intuition that serves as a foundation for knowledge. In brief, this supposition amounts to committing oneself to the idea we have already shown to be problematic that mere impressions ('bare presences', as McDowell often writes[14]) can constitute a justification for our empirical judgements. On the other hand, Coherentism is the position (which McDowell associates with Donald Davidson[15]) according to which experience, as an extra-conceptual impact on sensibility, cannot have any justificatory role for our judgements or beliefs. Davidson holds, indeed, that 'nothing can count as a reason for holding a belief except another belief'.[16]

As a response to the worry about the possibility of our thought's being in touch with reality, the Myth of the Given offers a position according to which we can 'acknowledge an external constraint on our freedom to deploy our empirical concepts' by grounding empirical justifications on impingements on the conceptual realm from outside.[17] As

12   Ibid., xii–xiii.
13   Sellars, 'Empiricism and the Philosophy of Mind,' esp. §7.
14   McDowell, *Mind and World*, 20 and *passim*.
15   Donald Davidson, 'A Coherence Theory of Truth and Knowledge,' reprinted in Ernest LePore (ed.), *Truth and Interpretation: Perspectives on the Philosophy of Donald Davidson* (Oxford: Blackwell, 1986), 307–19.
16   Davidson, 'A Coherence Theory of Truth and Knowledge,' 310.
17   McDowell, *Mind and World*, 6.

McDowell puts it: 'The idea is that when we have exhausted all the available moves within the space of concepts, all the available moves from one conceptually organized item to another, there is still one more step we can take: namely, pointing to something that is simply received in experience.'[18] But this pointing beyond concepts cannot be articulated. The Myth of the Given conceives external constraints to the spontaneous exercises of concepts as brute impacts from the exterior, bare presences that are received in the conceptual sphere. As McDowell writes, 'The idea of the Given is the idea that the space of reasons, the space of justifications or warrants, extends more widely than the conceptual sphere.'[19] But, as we so far have seen, if constraints are to be conceived as justifications for our empirical thought, we need *rational* constraints, and mere presences cannot serve that purpose. Although we can conceive of the Given as constraining thought, we cannot truly conceive of it as granting justification, for justification implies conceptual relations. Even though 'this picture secures that we cannot be blamed for what happens at that outer boundary, and hence that we cannot be blamed for the inward influence of what happens there,'[20] we need more than the removal of blame. Having a good reason to hold a belief goes beyond having no choice but to hold the belief, and this picture only secures the latter. What makes mythical the Myth of the Given is that it 'offers exculpations where we wanted justifications.'[21]

The failure of the Myth of the Given in making experience intelligible as a tribunal is what prompts us to take seriously the other side of the seesaw: Coherentism. This is the position according to which if we want to do justice to the dichotomy of logical spaces, but still conceive experience as made up of impressions, we cannot think of it as a tribunal. Experience cannot stand in judgement over our empirical thinking, it can only be *causally* relevant to a subject's beliefs or judgement, and so we are forced to renounce the idea of rational constraints external to our thought. That is to say, if we do not want to fall into the Myth of the Given, we *must* recognize that experience cannot count as a reason for holding a belief. In this way, the Coherentist picture confines us within the sphere of thought, depriving us of the very possibility of being in touch with something outside it, in the relevant sense of having rational constraints by the world. As a result we recoil into the Myth of the Given, only to see again that it cannot help us in making sense of how our judgements bear on reality.

According to McDowell, to dismantle the oscillation means to unmask as illusory the appearance of a philosophical obligation to occupy one or other of those two positions. In order to do this, we need a non-dualistic picture of the encounter between mind and world. That is, we should not add a third *theory* of the relation between mind and world, but we should be able to find the space for rethinking the place of mind and thought *in* nature. As we will see below, this space is made available by the notion of 'second nature.'[22] Through this notion, McDowell aims to show us a way to a position without theories, and furthermore, with no need for a theory. This method is explicitly

[18] Ibid.
[19] Ibid., 7.
[20] Ibid., 8.
[21] Ibid.
[22] Ibid., 84–8 and *passim*.

related to Wittgenstein and to his idea of the nature of philosophical clarification. Before saying something about Wittgenstein's idea of clarification and how McDowell applies it through the notion of second nature, it may be useful to look at the way in which McDowell arrives at discussing the human–animal divide.

McDowell's way of showing that the Cartesian picture of a dualism between reason and nature is not necessary goes through a rethinking of the interaction between sensibility and rationality, or, to put it in Kantian terms, between receptivity and spontaneity. McDowell holds that it is essential to the operation of receptivity that our conceptual capacities are already involved in it. According to McDowell, avoiding the Myth of the Given, without denying that experience plays a normative role, requires us to conceive experience as conceptual. This means that we must be able to recognize that conceptual capacities are already drawn on *in* receptivity – when we take in aspects of the world both in the case of empirical knowledge and in that of inner experience, when we make judgements about our own perception, thoughts, sensations and the like. That experience is conceptual means that conceptual capacities are not exercised *on* an extra-conceptual deliverance of receptivity. At the same time, to say that experience is conceptual doesn't mean that we create it. McDowell admits that 'because experience is passive, the involvement of conceptual capacities in experience does not by itself provide a good fit for the idea of a faculty of spontaneity'.[23] But the crucial point is that:

> How one's experience represents things to be is not under one's control, but it is up to one whether one accepts the appearance or rejects it. Moreover, even if one considers judgements that register experience itself, which are already active in that minimal sense, we must acknowledge that the capacity to use concepts in those judgements is not self-standing; it cannot be in place independently of a capacity to use the same concepts outside that context.[24]

This means that we are able to judge, for example, that something is white only because we are already able to move into the space of reasons, because we have a 'background understanding that includes, for instance, the concept of visible surface of objects and the concept of suitable conditions for telling what colour something is by looking at it'.[25] In the case of perception, McDowell's claim that the content of experience is conceptual amounts to the claim that perceptual episodes occurring in a human life are of a kind such that they must themselves be understood in terms that imply the power to reason about them; more specifically, it implies the power to reason about the relevance of such episodes to the person's life itself. This is the way in which experience (and perception, as an important aspect of the human condition) matters to us, the way in which we are open to the world.

[23] Ibid., 10–11.
[24] Ibid., 11.
[25] Ibid., 12.

I think that to understand this point is a good way to make sense of, or give content to, the idea of second nature. McDowell introduces this notion in order to make available a perspective from which we are not forced anymore to see our rationality as detached from our animal being. This is intended as the achievement of a position from which we can look at reason in terms of something natural. The notion of a human second nature, understood as a non-theoretical notion, is the reconciling reminder that we are animals whose natural being is permeated and transformed through and through by rationality. McDowell explains this point as follows:

> [H]uman beings are intelligibly initiated into . . . the space of reasons by ethical upbringing, which instils the appropriate shape into their lives. The resulting habits of thought and action are second nature.[26]

This means that through the idea of second nature, we can say that our rationality is part of (our) nature, and that our nature is the way it is because of a special relation between the biological features or potentialities we were born with and our upbringing, or *Bildung*. This special relation connecting what we may call our first and our second nature is, again, not to be understood as the exercise of conceptual capacities on an extra-conceptual Given; the relation is such that what we perceive (and desire, feel or experience in general) is already and inextricably informed by the presence of language, culture and rationality. In a word, we have inherited a *tradition*.

So it is not that we share the same perceptual system with animals, on which rationality simply sits. In McDowell's terminology, this would amount to having a 'highest common factor' conception of perception,[27] which is a version of the Myth of the Given. In appealing to second nature we are able to spell out and make intelligible what is unique about human beings as opposed to other animals, the *difference* between *us* and *them*, by reminding ourselves that our nature is rational in that peculiar way.

We have so far seen that in *Mind and World* the discussion about humans and animals is introduced as an illustration of an idea of conceptuality that could resist a conception of mind and experience informed by the mythology of givenness. To speak as McDowell does of *the* difference between humans and animals means that the distinction at issue points to a difference in kind concerning what we take to be unique about the human condition. When McDowell writes, for instance, that '[t]he spontaneity of the understanding cannot be captured in terms that are apt for describing nature . . ., but even so it can permeate actualizations of our animal nature',[28] part of what he is urging us to recognize is that the difference between humans and animals is a conceptual distinction and not something we discover through empirical observations. But what is the point of saying that the notion of the difference is conceptual? And where should we look to see in a perspicuous way the conceptual connections that we inherited as constitutive of our *Bildung* and that articulate the

---

[26] Ibid., 84.
[27] Ibid., 113.
[28] Ibid., 109.

idea of this difference? An answer to these questions may come from a comparison between McDowell's and Cora Diamond's accounts of the difference between humans and animals.

2. McDowell and Diamond on the difference between humans and animals

Here is a passage from Cora Diamond's paper 'Eating Meat and Eating People':[29]

> The difference between human beings and animals is not to be discovered by studies of Washoe or the activities of dolphins. It is not that sort of study or ethology or evolutionary theory that is going to tell us the difference between us and animals: the difference is . . . a central concept for human life and is more an object of contemplation than observation. . . . One source of confusion here is that we fail to distinguish between 'the difference between animals and people' and 'the differences between animals and people'. . . . In the case of the difference between animals and people, it is clear that we form the idea of this difference, create the concept of the difference, knowing perfectly well the overwhelmingly obvious similarities.[30]

In this passage Diamond says that any discovery of similarities between humans and animals with respect to properties or capacities that we might have previously considered as necessary conditions for being a person (such as the capacity to communicate) could not diminish our sense of *the* difference between us and them. Her point is directed against such authors as Peter Singer and Tom Regan who think that any appeal to a fundamental difference between human beings and animals amounts to a form of 'speciesism'.[31] In maintaining the difference, both Diamond's and McDowell's discussions could be read as presenting radical criticisms of theoretical positions (of which Singer's and Regan's are exemplary) that tacitly assume that the concepts of human beings and animals we draw on in moral thought are morally neutral, biological concepts.[32] But I will not be concerned with this criticism here. Instead, I want to focus on the difference between McDowell's and Diamond's accounts of human specialness.

---

[29] Cora Diamond, 'Eating Meat and Eating People,' in *The Realistic Spirit: Wittgenstein, Philosophy, and the Mind* (Cambridge, MA: MIT Press, 1991), 319–34.

[30] Ibid., 324.

[31] For the background of this debate see the classics by Peter Singer, *Animal Liberation: A New Ethics for our Treatment of Animals* (New York: Random House, 1975), and Tom Regan, *The Case for Animal Rights* (Berkeley, CA: University of California Press, 1983).

[32] For a description along these lines of Singer's and Regan's work on animal rights see Alice Crary, 'Humans, Animals, Right and Wrong,' in *Wittgenstein and the Moral Life: Essays in Honor of Cora Diamond* (Cambridge, MA: MIT Press, 2007), 381–404. For Diamond's criticism see also Cora Diamond, 'The Importance of Being Human,' in *Human Beings*, ed. David Cockburn (Cambridge: Cambridge University Press, 1991), 35–62.

McDowell's interest in the human–animal divide is prompted by the threat of a certain intellectual embarrassment in connection with the subject of animals; since his picture of our being open to the world is a consequence of our ability to engage in the conceptually articulated thought that records how things are in that world, he is led to an austere position about other animals' mindedness. For, if 'the objective world is present only to a self-conscious subject, a subject who can ascribe experiences to herself' and 'it is the spontaneity of the understanding, the power of conceptual thinking, that brings both the world and the self into view', it follows that '[c]reatures without conceptual capacities lack self-consciousness and . . . experience of objective reality'.[33] So, in McDowell's view, we *must* recognize that mere animals have only a kind of proto-subjectivity, and not the full-fledged subjectivity which is the exclusive prerogative of human beings. Even if animals are not Cartesian *automata*, their life is 'structured exclusively by immediate biological imperatives', and it is 'shaped by goals whose control of the animal's behaviour at a given moment is an immediate outcome of biological forces'.[34]

It seems that McDowell is prompted to speak about animals only by a theoretical interest in defending his picture of human nature as essentially second nature against the idea of non-conceptual experience. He focuses on animal mindedness, considering it as nothing more than a hard case, to test the validity of his picture of rationality. This argumentative requirement seems to drive him to a traditional task of constructive philosophy – that of demarcating the limits of subjectivity – that is at odds with the nature and purpose of his own conception of philosophical method. In fact, for what could be called a 'quietistic' account of rationality and of the human–animal divide it should be enough to stop at the point at which we are led to see the intelligibility of a position outside the oscillation, and that is obtained by picturing human nature as essentially a *second* nature.[35]

In the quoted passages McDowell is shifting his perspective, making a non-innocent move from a therapeutic[36] use of the notion of second nature to what might seem the basis of a substantive philosophical thesis about human rationality. For it is one thing to offer a reconciling picture of human mindedness pointing to the fact that 'we have what mere animals have, perceptual sensitivity to features of our environment, but we have it in a special form',[37] in order to show that it is not necessary for us to see human mind as something detached from nature; it is a different thing to say that animal lives are structured exclusively by immediate biological imperatives, or that they have a proto-subjectivity, and so on, as the result of a transcendental argument about human conceptuality. By tracing *in advance* the boundaries of what can be said to be 'properly human',[38] McDowell fosters a sense of strangeness in thinking of reason as part of nature that is at odds with his own quietistic aspiration. My criticism is thus concerned

[33] McDowell, *Mind and World*, 114.
[34] Ibid., 115.
[35] For my use of the term 'quietism', see footnote 2.
[36] Again, see footnote 2.
[37] McDowell, *Mind and World*, 64.
[38] See for instance *Mind and World*, 117, where McDowell speaks of a 'properly human life'.

with this Kantian inkling in McDowell's exposition, a transcendental element of a Kant not read in the light of Wittgenstein.

Diamond's way of approaching the idea of a fundamental difference between human beings and animals is different in its spirit. She says that:

> A difference like that may indeed start out as a biological difference, but it becomes something for human thought through being taken up and made something *of* – by generations of human beings, in their practices, their art, their literature, their religion, their ethics. . . . It is absurd to think that these are questions you should try to answer in some sort of totally general terms, quite independently of seeing what particular human sense people have *actually* made out of the differences or similarities you are concerned with. And this is not predictable.[39]

Again, Diamond is speaking against utilitarian accounts of animal ethics. But it is possible to read her words as a correction, or as an internal or sympathetic criticism, of an aspect of McDowell's Kantianism. From her point of view, the concept of the difference we may find between ourselves and animals is part of our life with them, and not the product of a transcendental argument about human conceptuality. The reality of the difference lies, for instance, in the fact that we cannot seriously write a biography of a dog, or in the fact that one of the activities through which we, or many of us, come to learn the concept of being human is that of sitting at a table and eating meat. In general, we create the concept of the difference through the variety of our responses to animal life, and *that* should be the starting point for a realistic philosophical account of human specialness.

So my point is that McDowell's discussion about the difference in kind between humans and animals does not accord with his own conception of how concepts work. What I take to be a general worry about McDowell's treatment of human rationality is that he tends not to be attentive to the details of our conceptual life. He offers a general account of the encounter of mind and world, and argues that in the relation between thought and reality the latter is given conceptually, and that we find our way in the world by being sensitive to the dictates of reason. But this tends to downplay the variety of ways in which we actually make sense of the world, the ways of our being sensitive to reality.

I want to focus on this point, and conclude with some brief remarks about McDowell's quietism and his Kantian way of proceeding. Recently McDowell has modified his view about the sense in which our sensory capacities are conceptual.[40] The crucial assumption that McDowell has rejected is that the actualization of conceptual capacities in experience, which is required for experience to stand within the space of reasons, means that experience must have propositionally articulated conceptual content. Thought of in this way, experience must have a *that-structure*, such that, when

---

[39]   Diamond, 'Eating Meat and Eating People,' 351.

[40]   See John McDowell, 'Avoiding the Myth of the Given,' in *Having the World in View* (Cambridge, MA: Harvard University Press, 2009), 256–72, and 'What Myth?' in *The Engaged Intellect*, 308–23.

one hears, say, a river, one hears that the river is ahead, and when one sees a book (which happens to be on a desk), one sees that there is a book on the desk. What is distinctive about the propositional content is not just its that-structure, but the fact that what it captures is the same as the content figuring in judgements or beliefs registering that things are thus and so. It is this idea that McDowell now questions, rather than the intuitively plausible notion that we can experience that something is the case. That is, experience is still conceived as conceptual, but now McDowell acknowledges a more complex idea of what it means for us to have an experience and, related to this, that the exercise of our conceptual capacities is not reducible to mere judgement making. For if experience, on which judgements are made, has a conceptual shape, which particular conceptual shape does it have? In order to answer this question, we should look at the actual experiences we may have in particular situations. This is the reason why McDowell now grants that experience is not always discursively conceptually shaped. This picture of conceptuality requires a *look and see* strategy that is able to show the articulation of the meanings we are responsive to. In the case of our idea of human specialness, this, again, would amount to addressing the question 'what human beings have *made of* the difference between human beings and animals'.[41]

3. McDowell and the big questions

In the light of this latter reminder, it is possible to state the kind of criticism I proposed by saying that McDowell does not give content to the idea that the human–animal divide may be something deeply rooted in our life, and that is because he takes it to be just a hard case to test the validity of his own answer to the Big Question about the general encounter of mind and world. This point about the big philosophical questions brings us to the issue of philosophical clarification, as Wittgenstein understood it. In her paper 'Criss-Cross Philosophy',[42] Diamond is concerned with Wittgenstein's criticism of his early conception of philosophical method. Diamond characterizes the difference between Wittgenstein's early and later method as one regarding the ability to practice philosophical clarifications 'not in the shadow of a big question'.[43] From Wittgenstein's later point of view, the *Tractatus* presents a piecemeal approach to philosophical problems which is still guided by the idea that in philosophy we face fundamental problems, such as the one about the essence of propositions. As Diamond says:

> [In the *Tractatus*] the search for the essence of language is, in theory, *überwunden*, overcome. But it is really still with us, in an ultimately unsatisfactory, unsatisfying, conception of what it is to clarify what we say.[44]

---

[41] Diamond, 'Eating Meat and Eating People,' 350–1.
[42] Cora Diamond, 'Criss-Cross Philosophy,' in Erich Ammereller and Eugen Fischer (eds), *Wittgenstein at Work* (London: Routledge, 2004), 201–20.
[43] Ibid., 218.
[44] Ibid., 207.

In order to escape the big questions, it is important to see philosophy as made up of particular problems. According to Diamond's reading, the *Philosophical Investigations* shows us the possibility of a piecemeal method that faces philosophical problems as 'standing on their own, to be dealt with on their own'.[45] And it accomplishes this task by, for instance, offering different kinds of language games, that is, imagining particular and concrete situations. This is part of the reason why in Wittgenstein's writings we do *not* find anything like this: 'Since beliefs are intelligible only as situated in the weave of our life, it is a mistake to think of a dog as having any'.

Of course the fact that Wittgenstein would have disagreed with McDowell about how to practice philosophical clarifications 'not in the shadow of a big question' does not prove McDowell wrong. Nonetheless, it shows a difficulty in combining Wittgenstein's conception of philosophy with the philosophical tradition of the Big Questions which is still present in *Mind and World*.[46] This difficulty is revealed in McDowell's treatment of the difference between humans and animals: the distinction is introduced to finally relieve us of the pressure of a conception of nature that forces us to think of human mindedness as something mysterious, but it immediately drives McDowell to trace in advance the boundaries of subjectivity. In so far as McDowell commits himself to Wittgenstein's idea that philosophy 'leaves everything as it is', his severe and rather aprioristic demarcation between human and animal realms extends beyond the scope of this commitment, and becomes something more than the 'reminder' that 'we have what mere animals have, perceptual sensitivity to features of our environment, but we have it in a special form'.

In this essay I have presented McDowell's account of human specialness in terms of a fundamental difference between humans and animals. I have then shown that the concept of that difference is relevant for a Wittgensteinian understanding of the

---

[45]   Ibid., 210. Diamond clarifies the way in which, according to her reading of Wittgenstein, philosophical problems are to be faced as 'standing on their own' as follows: '(1) . . . I don't mean that Wittgenstein thought of philosophical problems as having no light to shed on each other. He very definitely thought they did shed light on each other; but the question is how they do so. It is very natural for us to organize our thoughts about the connection of problems to each other through the idea of something both deeper and more general than the individual problems themselves. This conception is present in the *Tractatus* and survives the supposed *Überwindung* of the questions. Some or other version of it characterizes many readings of the *Investigations*, for example, that of Kripke, who sees the treatment of particular problems as falling out of a general treatment of Big Questions about Meaning. (2) In saying that Wittgenstein was not, in the *Investigations*, concerned to provide a new answer to the Big Question of the nature of language, I don't mean that he was not concerned with the nature of language. I mean that there is a difference between seeing such a question as a Big Question and seeing it simply as a problem or rather a group of problems, philosophical problems that can be approached through the methods he had developed' (ibid.).

[46]   Michael Friedman raises a similar worry, even if in the context of a hasty reading of McDowell's *Mind and World*. He writes: 'In light of the historical-philosophical tangles produced by McDowell's attempt to bring Wittgensteinian "quietism" into some kind of explicit relation with the philosophical tradition nonetheless, one can only conclude, in the end, that Wittgensteinian quietism may itself only make sense in the context of Wittgensteinian philosophical method' (Michael Friedman, 'Exorcising the Philosophical Tradition,' in *Reading McDowell: On 'Mind and World,'* ed. Nicholas H. Smith (London: Routledge, 2002), 25–57, at 48).

place of rationality in nature. I then developed an internal criticism of McDowell's transcendental way of approaching this topic by using Diamond's insights about the importance of the details for a realistic philosophical account of human mindedness. My aim has been to give support to the view that the difference between humans and animals, between *us* and *them*, is constitutive of our understanding of what it means to be humans, but this is not something we can explain in advance of looking at the weave of our life with them.[47]

[47] I want to thank Nora Hämäläinen, Niklas Forsberg and Mikel Burley for helpful comments on previous drafts and for their forbearance as editors. This essay also benefited from conversations with Alice Crary, Zed Adams and Piergiorgio Donatelli.

11

Wittgenstein, Wonder and Attention to Animals

Mikel Burley

[Wittgenstein] told me that he had got to know some wonderful characters in Norway. A woman who had said to him how fond she was of rats! 'They had such wonderful eyes.' This same woman once sat up every night for a month waiting for a sow to farrow, so as to be on hand to help if necessary. This attention to animals seemed to have pleased Wittgenstein especially.[1]

The phrase 'attention to animals' can be used in different senses, some of which have a more ethical emphasis than others. When Maurice Drury writes of Wittgenstein that he seemed to be especially pleased by the Norwegian woman's attention to animals, it seems fair to assume that one of the factors that Wittgenstein found pleasing was the ethical attitude that the woman displayed – the fact that she *cared* about animals. She cared about the sow enough to sit up at night in case it needed help when giving birth. And in the case of rats, too, the woman's attitude seems to have been one of caring, although it is possible that her fondness took an aesthetic form at least as much as an ethical one. To find the eyes of rats wonderful (or fascinating or intriguing) is not necessarily to adopt a particular ethical stance towards them; but one might suppose that forms of appreciation such as these at least make available a kind of ethical stance that may otherwise remain precluded. If, for example, a laboratory technician, whose job is to inject rats with a toxic chemical, begins to wonder at their eyes and behaviour – to see them as wonderful – she may then be liable to encounter ethical difficulties with the task she is expected to perform.

This essay will explore the notion of attention to animals in the light of Wittgenstein's philosophy. In particular I want to bring out some of the ways in which closely observing animal behaviour can generate a sense of wonder. Sometimes this may be wonder at a likeness, a similarity, between the animal and oneself, or between the animal and human beings more generally; at other times it may be wonder at the utter strangeness of the animal. There is perhaps a further sense of wonder to be had through the very awareness of this tension in the range of our possible responses: at the fact that animals which at certain times appear thoroughly familiar to us, at other times may display

[1] M. O'C. Drury, 'Conversations with Wittgenstein,' in *Recollections of Wittgenstein*, ed. Rush Rhees (Oxford: Oxford University Press, 1984), 120.

behaviour which strikes us as alien and unintelligible.[2] Moreover, an awareness of the mysteriousness of the behaviour of other species can sometimes engender reflection upon the mysteriousness of much of what *we* – human beings – do; and this in turn may contribute to what Cora Diamond has termed 'the feeling of solidarity in mysterious origin and uncertain fate' that may obtain between us and other people or animals.[3] Some elaboration of this thought will occur in the latter part of the essay. I want to begin, however, with an anecdote from my own experience, to which I shall relate the notion of *wondering at the world* expressed in Wittgenstein's 'Lecture on Ethics'.[4]

1. Wondering at animals and wondering at the world

On a summer's afternoon several years ago, I was sitting outdoors reading a book when I heard a frantic high-pitched buzzing sound, like the sound that a fly or bee makes when trapped inside a confined space such as a plastic bag. There was a large upturned plant pot nearby, so I suspected that a buzzing insect had become trapped underneath. But as I approached the pot, I noticed that the buzzing was coming not from beneath but from a little to one side of it. There I saw two wasps, one of which had the other pinned to the ground on its back. The attack was evidently ferocious, and my instinct was to intervene. Picking up a long thin twig, I moved one end of it towards the wasps. Upon my carefully prodding the dominant wasp, however, it did not fly off as I'd expected it to, but persisted in its vigorous assault. As I continued to watch, I noticed that this wasp was devouring its victim. It chomped into the other's thorax and abdomen, consuming these parts of the other's body at rapid speed. The pace was such that it appeared to me somewhat as though the scene had been filmed and then sped up, as when one presses the fast-forward button on a DVD player. After less than a minute the dominant wasp did fly off, leaving the other – consisting now of a mere head, legs and the back of its exoskeleton – crawling feebly, and now silently (the remains of its wings being in no condition to vibrate), across the concrete slab where the fatal attack had occurred. I placed a leaf over its dying body, positioned my sandaled foot over the leaf, and trod down hard to finish off the wretched creature as quickly as I could. I thought it had suffered more than enough already.

Unsurprisingly, this was not the first or the last time that I have witnessed one living creature violently attacking another. In my days as a hunt saboteur, for example, I came close enough to the end of some hunts to see foxes or mink being ripped to pieces by packs of half-starved hounds. Yet there was something

---

[2] Cf. Cora Diamond's remark on the 'sense of astonishment and incomprehension that there should be beings so like us, so unlike us, so astonishingly capable of being companions of ours and so unfathomably distant' ('The Difficulty of Reality and the Difficulty of Philosophy,' in Stanley Cavell et al., *Philosophy and Animal Life* (New York: Columbia University Press, 2008), 61).

[3] See Cora Diamond, 'The Importance of Being Human,' in *Human Beings*, ed. David Cockburn (Cambridge: Cambridge University Press, 1991), 55.

[4] Ludwig Wittgenstein, 'A Lecture on Ethics,' *Philosophical Review*, 74 (1965), 1–12.

peculiarly arresting and memorable about the wasp incident. In the case of the fox-
and mink-hunts, there had at least been some intelligible context to the killing,
however morally disagreeable that context might be. There had also been a clear
desire on the part of myself and of my fellow saboteurs to prevent the killing
taking place, and a consequent experience of frustration and failure when our
preventative tactics were unsuccessful. But in the case of the wasps, I was (despite
my tentative attempts at intervention) far more detached from the incident. There
was something about the sight of one wasp voraciously devouring the other that
made me feel strangely distanced from their world. As I hinted earlier, it was a little
like watching a film – a science-fiction movie in which two alien beings are battling
one another – and yet, unlike in the case of watching a movie, knowing that this
was *real*. Real, but unintelligible to me. I was amazed and in awe of the furious
violence manifesting before my eyes. No doubt, in some ways, such activity would
appear less mysterious – or more familiar at any rate – to a trained entomologist;
it might be the kind of scene they have witnessed many times, and they may have
devised theories about why such behaviour takes place. But all the same, from my
point of view, it is difficult to imagine how one could ever cease to be amazed that
such things occur.

The chapter entitled 'Brute Neighbors' in Henry David Thoreau's *Walden* contains
a description of an experience with some similarities to mine.[5] Writing of having
witnessed 'a war between two races of ants', Thoreau readily compares the scene to
human combat: 'I was myself excited somewhat even as if they had been men. The
more you think of it, the less the difference.' In my own case, I felt unable to relate
the cannibalizing action of the wasp to human activity. But perhaps, upon further
reflection, it may not seem so divergent from what some humans have been known to
do, especially under conditions of war.

When Wittgenstein, in his 'Lecture on Ethics', speaks of 'wondering at the existence
of the world' and of 'seeing the world as a miracle',[6] I presume that wasps or ants
fighting to the death is not the sort of thing that is at the forefront of his mind. But
for me, at least, it is experiences such as observing the frenzied wasp-fight that bring
home what I take Wittgenstein to have meant.

For Kant, famously, it was 'the starry heavens above [us] and the moral law within
[us]' that 'fill the mind with ever new and increasing admiration and awe, the oftener
and more steadily we reflect on them';[7] but it is not inconceivable that *any* phenomenon
could play this role for a given individual. To see something as a miracle, and hence
to wonder at it in what I take to be Wittgenstein's sense, is not to see a particular kind
of *thing*, but rather to see it in a particular kind of *way*. It is to view something, as
Wittgenstein was apt to express it in his early notebooks, *sub specie aeternitatis*. 'The

[5] See, for example, Henry D. Thoreau, *Walden: An Annotated Edition*, ed. Walter Harding (New York:
Houghton Mifflin, 1995), 223–6. I am grateful to Tom Kettunen (Åbo Akademi University) for
reminding me of this chapter.
[6] Wittgenstein, 'A Lecture on Ethics', 8 and 11.
[7] Immanuel Kant, *Critique of Practical Reason* [1788], trans. Lewis White Beck (Indianapolis:
Bobbs-Merrill, 1956), 5:161.

usual way of looking at things', he writes, 'sees objects as it were from the midst of them, the view *sub specie aeternitatis* from outside'.[8] I certainly felt outside and disconnected from the world of the wasps, and this disconnectedness was reinforced by the dominant wasp's obliviousness to my poking it with a stick. Excluded from their world, I was a mere observer, spectating from outside. And yet, in another (perhaps platitudinous) sense, I knew full well that we shared the same world; and it was in part this tension, between the alienness of the scene I was witnessing and the knowledge that this was an everyday feature of the world we inhabit, that contributed towards my peculiar feeling of astonishment.

None of this is meant to deny that there are features of the universe that are more characteristically likely than others to inspire awe and wonder in us, and a feeling of the sublime. It is no surprise that Kant picks out the starry heavens as one such feature. Perhaps more surprising is that this should be paired with 'the moral law within'. But if Kant means by this that we can feel awestruck by the sheer existence of morality – which operates in the world and yet seems to have its grounding in something beyond the world – then I can see why he says it.[9] Nevertheless, whatever exactly Kant meant by what he said, I would want to affirm that we can follow William Blake in having the same sort of response to a grain of sand as to the world as a whole, and to a wild flower as to the starry heavens above.[10] I would want also to echo the observation made by Wittgenstein and Rush Rhees, that terrible and destructive phenomena can be celebrated, in the sense of being ceremonially marked, as well as auspicious and beneficial ones. Taking his cue from some of Wittgenstein's remarks on Frazer's *Golden Bough*, Rhees urges us to recognize that rites in which, for example, human figures, even children, are burnt in effigy, need not be construed instrumentally, 'as methods designed to *ward off* the terrible things they celebrate', but may be seen as sacraments, as expressions of wonder at – and hence as acknowledgements of the importance of – the terrible phenomena themselves.[11] It would be a misrepresentation

---

[8] Ludwig Wittgenstein, *Notebooks, 1914–1916*, trans. G. E. M. Anscombe, 2nd edn (Oxford: Blackwell, 1979), 83e.

[9] Cf. M. O'C. Drury, *The Danger of Words* (London: Routledge and Kegan Paul, 1973), 128: 'Oughtness is as much a datum of consciousness as the starry vault above. Both should continue to fill us with constant amazement.'

[10] See Blake's poem *Auguries of Innocence* (1803), first stanza: 'To see a World in a Grain of Sand | And a Heaven in a Wild Flower | Hold Infinity in the palm of your hand | And Eternity in an hour | . . .' (in *The Complete Poetry and Prose of William Blake*, ed. David V. Erdman (Berkeley, CA: University of California Press, 2008), 490). Cf. Wittgenstein's remark that 'no phenomenon is in itself particularly mysterious, but any of them can become so to us, and the characteristic feature of the awakening mind of man is precisely the fact that a phenomenon comes to have meaning for him' ('Remarks on Frazer's *Golden Bough*', in *Philosophical Occasions, 1912–1951* (Indianapolis: Hackett, 1993), 115–55, at 129).

[11] Rush Rhees, 'The Fundamental Problems of Philosophy', *Philosophical Investigations*, 17 (1994), 573–86, at 578–9. Cf. D. Z. Phillips, *Recovering Religious Concepts: Closing Epistemic Divides* (Basingstoke: Macmillan, 2000), 43. Among the relevant remarks of Wittgenstein's is one in which he asks 'how is it that in general human sacrifice is so deep and sinister?' and then addresses the question in a way that suggests that seeing it as deep and sinister is a primitive response on our part: 'the deep and sinister do not become apparent merely by our coming to know the history of the external action, rather it is *we* who ascribe them from an inner experience' ('Remarks on Frazer's *Golden Bough*', in *Philosophical Occasions*, 147).

to say of Wittgenstein and Rhees that they are suggesting that rites can celebrate terrible phenomena in the sense of celebrating those phenomena as *good* and not really terrible after all. The point is not that the rites try to transform something terrible into something benign. Rather, it is that the rites celebrate that which is terrible *as terrible* – they acknowledge that terror is part of the world and part of our lives. The particular form that the acknowledgement takes will, of course, depend on the ritual itself.[12]

2. Expressing an intention

Before moving on from the wasp incident I have described, I want to make a few further remarks that connect it with a point of Wittgenstein's. Wittgenstein proposes in the *Investigations* that 'the natural expression of an intention' is to be seen in 'a cat when it stalks a bird; or a beast when it wants to escape'.[13] This proposal has been criticized by Elizabeth Anscombe on the grounds that the *expression* of an intention requires language, or at least the deployment of conventional bodily gestures or movements, which constitute parts of the communicative repertoire only of language-using creatures, namely ourselves.[14] I take Wittgenstein's point to be that the cat's behaviour vividly embodies a direction or orientation towards a particular act; its whole demeanour is oriented towards pouncing and capturing the bird. Anscombe may not have demurred from *this* way of describing the cat's behaviour. It is just that she would not have agreed that the cat's demeanour – its 'crouching and slinking along with its eye fixed on the bird and its whiskers twitching', as Anscombe herself evocatively puts it[15] – amounts to the expression of an intention.

My view on this disagreement between Anscombe and Wittgenstein is that it turns on an ambiguity in the term 'expression' as it is being used in this context. Clearly the cat does not have the capacity to express its intention in the sense of *telling* us what its intention is, or even by articulating its intention to us by means of gestural or sign language. We might put this by saying that the cat cannot *voluntarily* express its intention. But it can – if we admit the propriety of Wittgenstein's use of 'expression' (*Ausdruck*) – *exhibit* or *display* its intention to us through, or by means of, its behaviour. Anscombe does not want to deny that the cat's behaviour is intentional; she denies merely that the cat's intention can be *expressed* by the cat. But I'm with Wittgenstein on this matter, for it strikes me that there is a legitimate place for our describing the cat's intention as being expressed in or through its behaviour, not verbally but bodily.

---

[12] One form that it may take is exemplified within certain Hindu traditions, wherein we find the goddess Kālī being worshipped both as the divine mother and sometimes, apparently simultaneously, as a ferocious destructress who demands the sacrifice of animals and perhaps of human beings as well. See, for example, Elizabeth U. Harding, *Kali: The Black Goddess of Dakshineswar* (York Beach, ME: Nicolas-Hays, 1993), esp. 117–19; R. C. Muirhead-Thomson, *Assam Valley: Beliefs and Customs of the Assamese Hindus* (London: Luzac, 1948), esp. 32.

[13] Ludwig Wittgenstein, *Philosophical Investigations*, 3rd edn (Oxford: Blackwell, 2001), §647.

[14] G. E. M. Anscombe, *Intention*, 2nd edn (Cambridge, MA: Harvard University Press, 2000), 5.

[15] Ibid., 86.

It is less clear what Wittgenstein has in mind with his reference to a beast that wants to escape, partly because he does not tell us what sort of beast he is thinking of, and partly because the beast's behaviour is not described. Perhaps if one imagined, say, an imprisoned chimpanzee throwing itself against the bars of its cage, this would exemplify for Wittgenstein an intention to escape – although it may just as well be described as exemplifying extreme rage or frustration (though these latter descriptions need not, of course, be inconsistent with the description of the behaviour as expressing, or exemplifying, an intention to escape). But it is not crucial that we tidy up the details of a case such as this in order to see that Wittgenstein's remarks on 'the natural expression of an intention' may prompt us to contemplate the wasp incident in a particular way. Indeed, I did not require any prompting from Wittgenstein in order to see the attack and act of devouring as an embodiment of concentrated aggression and violence.

However, whether I would want to attribute to the victorious wasp an intention to kill is doubtful; at least, I feel some difficulty in making this attribution. Part of this difficulty derives from the fact that the act of killing was so singularly directed, so intense and efficient, that there was, as it were, no *room* for the attribution of an attitude of intending or anything else, just as, in at least many instances, there is no room for attributing attitudes to machines.[16] We may, of course, attribute attitudes to machines in a figurative sense, such as when we speak of a computer as 'cranky' or 'bad tempered', and we may on occasions (especially if we have children or are children ourselves) enter into games of make-believe in which attitudes are attributed to all sorts of inanimate objects, including machines. But, in what I am inclined to call the primary sense of 'attitude', machines cannot have them – they simply perform, automatically, the task that we have set for them (or, on certain occasions, fail to perform it, but their failure is not the sort that would warrant an accusation of guilt).

Maybe I have been conditioned by too many horror and science-fiction movies featuring giant insects into supposing that they are machine-like in their behaviour; but there *is* something mechanical about them. When we say of a human being – an experienced and well-trained assassin for example – that he is a killing machine, we mean that he is an efficient killer, and perhaps that his emotional detachment is such that he will not be distracted from his task by moral qualms; but the term 'machine' is to be taken metaphorically. In the case of the wasp that I saw devouring its victim, however, the expression 'killing machine' would not be so metaphorical – or would, at least, be metaphorical in a different way. Admittedly, there was nothing cool or detached about the wasp's behaviour; indeed, as I have suggested, in the act of killing it could be described as an embodiment of aggression. And in seeing the wasp under this aspect, it is difficult to resist the thought that one is attributing to it an emotional state, which pulls against the characterization of its behaviour as mechanical.

I don't think there is any neat solution to this descriptive difficulty. As Wittgenstein notes, when we see a wriggling fly, the attribution of pain 'seems to get a foothold', which

[16]   In conversation, Sue Richardson has suggested that in describing the wasp incident it would be correct to say that the dominant wasp exhibited a '*focused intent* to kill' (as opposed to an *intention* to kill). But I am not so ready to separate the terms 'intent' and 'intention' in this context.

is completely absent in the case of stones.[17] Perhaps, too, certain emotional predicates – 'angry', 'enraged', 'furious', 'murderous' – get a foothold when we observe one wasp voraciously consuming another. But still, I would not want to describe the aggression as expressing an intention or a desire to kill; it was instinctive without involving deliberation. Of course, one can say of a cat that its stalking of a bird is instinctual, and yet – if Wittgenstein's thought is on target – that it expresses an intention nevertheless. This is indicative of the semantic range of the word 'instinct' (and the concept of an instinct): in some cases it is bound up with intention and in others not. In the case of the wasp, I am proposing that it is not.

Now, with these matters concerning intention and instinct in the background, let us consider further the interrelations between strangeness and familiarity in our responses to animals.

3. Strangeness and familiarity

'[N]ow and again there appears a novel which opens up a new world not by revealing what is strange, but by revealing what is familiar.'[18] George Orwell had Joyce's *Ulysses* primarily in mind when he wrote these words. But drawing our attention to everyday things in a way that reveals them afresh is certainly a common feature of Wittgenstein's later work, and also of the work of many thinkers who have been influenced by his methods.[19] One such thinker is John Churchill, and in order to develop the idea that there may, on occasions, be a tension – or a reflexive interplay – between strangeness and familiarity in our encounters with animals, I want at this juncture to adduce one of Churchill's memorable examples.

In the recounted incident, Churchill was getting some exercise by running along a country road when he 'was joined by a large, unfamiliar dog who seemed to want to go along simply for the run.'[20] 'As we trotted along,' continues Churchill,

> we came upon two road-kills, an armadillo and a possum, lying within a few feet of each other. The dog stopped to sniff both, and I stopped to watch. The armadillo was freshly killed; not the possum. After an inspection of the former, the dog sniffed at the rancid remains of the possum with great eagerness. Half-turning, he rolled onto his back on top of the possum and wallowed, as dogs do, grinding himself against it. After a minute he got up, his fur thoroughly soaked and smeared, and we trotted on.[21]

---

[17]   See Wittgenstein, *Philosophical Investigations*, §284.

[18]   George Orwell, 'Inside the Whale,' in his *'England, Your England' and Other Essays* (London: Secker & Warburg, 1953), 96.

[19]   I am not the first to draw this connection between Orwell's remark and the work of Wittgenstein. See Frank Cioffi, 'Overviews: What Are They of and What Are They For?' in *Seeing Wittgenstein Anew*, ed. William Day and Victor J. Krebs (Cambridge: Cambridge University Press, 2010), 291–313, at 291.

[20]   John Churchill, 'If a Lion Could Talk . . .' *Philosophical Investigations*, 12 (1989), 308–24, at 321–2.

[21]   Ibid., 322.

In pondering this incident, Churchill notes that, although there are many forms of dog behaviour that we understand (at least to some extent), in this particular case he is strongly inclined to say that he doesn't know *what* the dog was doing, that the dog's 'behaviour is beyond the frontier of our kinship, beyond the point at which human and canine being in the world diverge'.[22] In reflecting further, Churchill considers whether there is anything analogous to the dog's behaviour in our human lives, and he observes that odours play important roles not only in the lives of animals such as dogs, and of 'so-called "primitive" peoples', but also 'in the interactions of Western middleclass people'.[23] 'Perhaps it is not only the behaviour of the dog I do not understand,' Churchill muses. 'What is the whole story of the role in my own life of sweat and of pheromones, of bathing and of soap, deodorants, and colognes? Do I have a true account of all that?' Then he connects these thoughts with those of Wittgenstein in the following poignant remark:

> It is characteristic of tracing out the implications of Wittgenstein's work that we discover that the reflections on what we *know* we don't understand brings into focus the fact that what we *do* understand in our own lives often rests on the unexplored ground of what we have simply accepted, without thought, as familiar and ordinary.[24]

Churchill makes an insightful observation here about a prevalent method deployed by Wittgenstein. The method involves noticing something that strikes us as strange in a phenomenon, then looking to see whether there is anything in our own lives that is relevantly similar to the phenomenon in question. If we manage to detect a similarity, then we may feel that we have learnt something about the phenomenon that we at first regarded as strange; and in a sense we *have* learnt something, namely that it is not so far removed from our own lives as we had assumed. In learning this, however, we at the same time come to see our own lives and practices in a new light. We come to see that they, too, have a certain strangeness about them, that although they may well share features with other phenomena in the world, they are not based upon solid and fully comprehensible foundations; rather, they are, as Wittgenstein puts it in *On Certainty*, 'held fast by what lies around [them]'[25] – by what surrounds them in the swirl or weave of our lives with others.

We can also see this method of Wittgenstein's at work in his *Remarks on Frazer's 'Golden Bough'*, where he seeks to disrupt Frazer's assumption that the magical practices of ancient or tribal peoples evince a pre-scientific conception of causal relations. In the case of the burning of effigies, for example, while Frazer takes this to be a misguided attempt to cause harm to the individuals represented by the effigies, Wittgenstein looks for analogies in a very different direction:

> Burning in effigy. Kissing the picture of one's beloved. That is *obviously not* based on the belief that it will have some specific effect on the object which the picture

---

[22] Ibid.
[23] Ibid.
[24] Ibid., 322–3.
[25] Ludwig Wittgenstein, *On Certainty* (Oxford: Blackwell, 1979), §144.

represents. It aims at satisfaction and achieves it. Or rather: it *aims* at nothing at all; we just behave this way and then we feel satisfied.[26]

In the final sentence of this remark we see that Wittgenstein is not trying to replace one explanation with another. He is trying to direct our attention away from the assumption that human practices must be based on beliefs or opinions which provide a rationale for the practices, and towards an alternative view, according to which there may be nothing upon which many of our practices are 'based'. A practice such as kissing a photograph of a loved one is not the sort of thing for which there can be a rationale; it is simply something we – or many of us – do. If we do it at all, we do it spontaneously, naturally. If someone were to ask us why we were doing it, the most natural thing to say would be that one loves the person in the photograph; and this is not to articulate a belief or opinion about the *effect* that one's action is likely to have. The extent to which someone sees the point in kissing a photograph will depend on the extent to which they share the impulse to do this themselves (in particular circumstances), or upon the extent to which they can relate this to relevantly similar impulses that they feel – the impulse to caress a loved one's garment when it is hanging in the wardrobe, for example.

It is, of course, entirely consistent with Wittgenstein's method of looking for analogies that in some cases there will be no relevantly similar analogue to be found in our own lives, and hence the gap of understanding is liable to remain unbridged, the activity in question remaining unintelligible to us. We may understand it enough to see that it is meaningful activity, and hence that there is *something to be understood*, without thereby understanding the meaning that it has. Indeed, Wittgenstein acknowledges 'that one human being can be a complete enigma to another' and that this can be learnt 'when we come into a strange country with entirely strange traditions'. In the case of such a people, Wittgenstein continues, 'Wir können uns nicht in sie finden.' 'We cannot find our feet with them,' as Anscombe translates this phrase;[27] or, keeping more closely to the German, 'We cannot find *ourselves* in them.'[28] It is in this context that Wittgenstein offers his gnomic and much-quoted remark that 'If a lion could talk, we could not understand him.'[29] And it is this remark that constitutes the starting point for Churchill's ruminations in the article from which I quoted earlier.

The point of Wittgenstein's 'lion' remark is, I take it, to emphasize in figurative terms the rootedness of mutual intelligibility in shared forms of life. There may be certain activities that we share with lions or other cats, such as stalking one's prey, which of course Wittgenstein himself mentions.[30] Even if we do not stalk prey ourselves, we know what it would mean for a human being to do so; and, for many of us, this knowledge will facilitate a more intimate understanding of at least *this* feature of feline behaviour

[26] Wittgenstein, 'Remarks on Frazer's *Golden Bough*,' in *Philosophical Occasions*, 123.
[27] Wittgenstein, *Philosophical Investigations*, p. 190.
[28] My emphasis. James C. Klagge translates the phrase this way in his *Wittgenstein in Exile* (Cambridge, MA: MIT Press, 2011), 43.
[29] Wittgenstein, *Philosophical Investigations*, p. 190.
[30] See ibid., §647, cited above.

than of the behaviour of, say, insects or fish. In the case of dogs there may be a great deal more that we have in common. One of the things that Churchill is pointing out in his reflections on the dog and the dead possum is that, at certain places, the attempt to see commonalities may fail, and yet that, in that very moment of failure, we may come to recognize that many features of our own lives are no less mysterious than those of a dog. What distinguishes much of our own behaviour from that of the dog is not that ours is supported by reasons whereas the dog's is purely instinctive and ungrounded; it is simply that, in the case of our behaviour, it is *what we do*. We understand it not in the sense of being able to explain it, or offer reasons that underpin it, but merely in the sense of being familiar with doing it. 'What has to be accepted, the given, is – so one could say – *forms of life*';[31] 'We can only *describe* and say, human life is like that.'[32] The understanding that is arrived at is not one that enables us to explain either the animal's form of life or our own; it is one that enables us to recognize that explanations have come to an end, bedrock has been reached.[33]

A nice example of a case where it seems plausible to say that mutual intelligibility does obtain between a dog and a human being is given by Simon Glendinning in his book *On Being with Others*. Glendinning recalls how he used to play a game with his mother's dog Sophie, in which he and Sophie would run around a small pond, his aim being to catch her and hers to avoid being caught.[34] On one occasion, the young Glendinning lost his footing on damp grass, and Sophie came up to him. 'I was unhurt,' writes Glendinning, 'but she licked my face anyway. / I do not see why this cannot be counted as a case of "mutual intelligibility". The dog could see my distress and I could see her sympathy.'[35] In cases such as this, we might – or Glendinning might, at any rate – want to say that if the dog could talk, we may well be able to understand her.

4. Attending to the voices of animals

When speaking of being able, or unable, to understand an animal that could talk, we should not overlook possible figurative senses of 'speaking' and 'hearing'. Cora Diamond draws attention to such senses when she writes that 'our *hearing* the moral appeal of an animal is our hearing it speak – as it were – the language of our fellow human beings'.[36] The point here is that terms such as 'hearing' and 'speaking' can take

---

[31]   Ibid., p. 192.
[32]   Ludwig Wittgenstein, *Remarks on Frazer's 'Golden Bough,'* trans. A. C. Miles, revised by Rush Rhees (Bishopstone: Brynmill Press, 1979), 3e.
[33]   'Explanations come to an end somewhere' (*Philosophical Investigations*, §1); 'If I have exhausted the justifications [*die Begründungen*] I have reached bedrock, and my spade is turned. Then I am inclined to say: "This is simply what I do" ' (ibid., §217). Cf. Ludwig Wittgenstein, *Lectures and Conversations on Aesthetics, Psychology and Religious Belief* (Oxford: Blackwell, 1966), 25: 'Why do we do this sort of thing? This is the sort of thing we do.'
[34]   The reverse of such a game – where the dog chases the human being – is amusingly portrayed in Buster Keaton's film *The Scarecrow* (1920).
[35]   Simon Glendinning, *On Being with Others: Heidegger–Derrida–Wittgenstein* (London: Routledge, 1998), 142.
[36]   Cora Diamond, 'Eating Meat and Eating People,' *Philosophy*, 53 (1978), 465–79, at 478.

on a moral sense, where to hear an animal speaking is to attend to it in a certain way, to perceive it as residing within the sphere of ethical concern. In much of Diamond's work on moral issues pertaining to animals she shows how an attitude of considerateness towards members of other species can stem, not from rational assessments of their biological or psychological similarities to typical human beings, but from a feeling of fellowship in mortality – a feeling or perception of the animal's being like us in its vulnerability to suffering, injury and death. Invoking a phrase from Joseph Conrad, Diamond speaks of 'the feeling of solidarity in mysterious origin and uncertain fate', which reaches out – or has the potential to reach out – to human beings afflicted by intellectual or bodily impairments, and also to 'the dead and the unborn', as well as to non-human animals.[37]

This moral responsiveness or attentiveness – the impulse to sit up all night with a heavily pregnant sow, for instance – is not something that can be acquired simply from reading the arguments of moral philosophers. Rather, Diamond emphasizes the importance of cultivating the moral imagination through reflection on literary examples – examples where concepts such as that of 'living creature' or 'fellow creature', or of a *community* of fellow creatures, are given voice.[38] In order to accept this contention, we need not dismiss the kinds of arguments put forward by moral philosophers as entirely worthless. We need instead to consider, for ourselves, whether cool comparative analysis of, for example, the cognitive or sensorial capacities of most adult humans vis-à-vis those of the average members of certain other species is likely to be sufficient for any significant change in moral sensibility to occur. It should also be noted that Diamond's emphasis upon the value of literature entails neither the redundancy of first-hand experience nor the *necessity* of intimacy with literature in order to develop or enhance moral capacities. It would be surprising, for example, if the Norwegian woman in Drury's anecdote developed her attentiveness to pigs by contemplating works of narrative fiction.[39] What I understand Diamond's suggestion to be is merely that a strong affinity obtains between a cultivated acquaintance with certain forms of literature – especially certain novels and poems – on the one hand, and a rich and expansive moral sense on the other.

While these considerations concerning the role of literature in our moral lives undoubtedly raise many issues that warrant further discussion elsewhere, the main point I want to bring out here is simply the resonance between these thoughts on moral attention and the themes that I have been exploring throughout this essay. These themes include that of the tension or interplay between strangeness and familiarity in our dealings with animals, and also that of wonder, both at the similarities and differences between animals and ourselves, and – further – at the unfathomable mysteriousness of the whole of life, this mystery consisting in the ultimate inexplicability of why we are

---

[37]   Diamond, 'The Importance of Being Human', 55. Cf. the Preface to Conrad's *The Nigger of the 'Narcissus'* (Garden City, NY: Doubleday, Page & Co., 1897), xiv.

[38]   See, for example, Diamond, 'Eating Meat and Eating People', 474.

[39]   A point well made by Olli Lagerspetz during discussion.

here or why things happen as they do.[40] In noticing the radical contingency and fragility of animal life we may come to see more vividly the extent to which our own lives, too, are fragile and contingent. Within this recognition of shared vulnerability may lie the source – or, at any rate, one important source – of ethical thought and action, and one way of articulating this recognition is to say that it consists in our coming to hear and acknowledge the voices of those with whom we are fellow inhabitants of this earth.

5. Concluding remarks

It has often been said that philosophy begins in wonder, and that, as Socrates proclaims in Plato's *Theaetetus*, wondering 'is characteristic of a philosopher'.[41] It might be assumed that philosophy's task is to facilitate a movement from wonder to knowledge, thereby replacing an attitude of awe with one of mastery and comprehension. But there is room for a conception of philosophy according to which it involves deepening rather than displacing our sense of wonder and astonishment. In this essay I have indicated a way in which reflecting upon our encounters with animals can provide an occasion for such a deepening to occur. The initial step consists in one's being struck by something in the animal – by a feature of its behaviour or its appearance. Sometimes this may not lead to further contemplation. There is no need to suppose, for example, that when the Norwegian woman in Drury's anecdote was struck by the appearance of rats' eyes this would necessarily lead to any further thought on her part concerning her relation to rats; rather, her seeing the eyes of the rats as 'wonderful' could be said to manifest an aspect of the relation in which she already stood. However, in other instances, the feature of the animal or its behaviour may generate a sense of wonder that is also a sense of incomprehension, which calls out for further contemplation (though not, or certainly not *necessarily*, an explanation).

In response to the dog rolling in the remains of a dead possum, Churchill was spurred to wonder why the dog was doing this, and to seek possible analogies in the behaviour of human beings. In this particular instance, despite his finding no analogue to illuminate the dog's behaviour, Churchill's attention was led towards features of human life whose enigmatic nature frequently goes unnoticed – features involving our own use of, and responses to, odoriferous substances and secretions. By allowing one's thoughts to move in directions of this sort it is conceivable that an initial reaction of disgust and utter bewilderment at an animal's behaviour could be transformed into a recognition of common fellowship, of what Diamond calls 'the feeling of solidarity in

---

[40] Cf. D. Z. Phillips' discussion of the 'sense of mystery' in his 'Wittgensteinianism: Logic, Reality, and God,' in *The Oxford Handbook of Philosophy of Religion*, ed. William J. Wainwright (Oxford: Oxford University Press, 2005), 447–71, at 460–1.

[41] Plato, *Theaetetus*, 155d: '. . . this is an experience which is characteristic of a philosopher, this wondering: this is where philosophy begins and nowhere else' (Plato, *Complete Works*, ed. John M. Cooper (Indianapolis: Hackett, 1997), 173). Cf. Aristotle, *Metaphysics*, 982b12–13: '. . . it is owing to their wonder that men both now begin and at first began to philosophize' (*A New Aristotle Reader*, ed. J. L. Ackrill (Oxford: Clarendon Press, 1987), 258).

mysterious origin and uncertain fate'; we come to acknowledge the strangeness within the very familiarity of our own lives, the ultimate absence of a rational ground for much of what we do as a matter of course. Although this can be an unsettling feeling, it can also open the way to enhanced moral responsiveness, as what had previously appeared to be an impermeable barrier between human and non-human worlds becomes less pronounced.

Also unsettling are those experiences, such as my witnessing of one wasp cannibalizing another, which elicit a terrible fascination. Part of this fascination consists in an awareness of how such incidents seem to embody something terrifying about the universe more generally – its destructiveness and obliviousness to the concerns of human beings, or indeed to the life and well-being of any other creature. In this way they can become occasions for wonder in a religious sense, an awe that stirs a profound humility, just as much as can gazing at the starry heavens.

So a deepening of wonder – and hence a deepening of one's experience of life – can occur through an increased sensitivity and attentiveness to animals; and Wittgenstein seems to have been alert to these possibilities. Without presenting any general theses concerning the lives of animals or the relations between animals and ourselves, nor seeking to instruct us on how animals ought to be regarded ethically, Wittgenstein provides poignant warnings against the urge to over-generalize, and reminders to pay close attention to the details of particular cases.[42] What I have suggested here is that our cultivating this attention to particulars is one of the ways through which our wonder at the endlessly variegated forms of life may deepen and increase.[43]

---

[42] See, for example, Wittgenstein's criticisms of the 'craving for generality' and the 'contemptuous attitude towards the particular case' in *The Blue and Brown Books: Preliminary Studies for the 'Philosophical Investigations'*, 2nd edn (Oxford: Blackwell, 1969), 17–19. See also his remark to Drury that his interest (in contrast with that of a philosopher such as Hegel) 'is in [showing] that things which look the same are really different' (Drury, 'Conversations with Wittgenstein,' 157).

[43] I am grateful to Sue Richardson for helpful conversations relating to the content of this essay, and to participants in the *Language, Ethics and Animal Life* conference at Uppsala University, 26–28 March 2010, for their comments and questions, especially Olli Lagerspetz. Thanks are due also to David Cockburn, Niklas Forsberg, Nora Hämäläinen and Tom Kettunen for written comments following the conference.

12

Honour, Dignity and the Realm of Meaning

Nora Hämäläinen

In this essay I will discuss the nature of Raimond Gaita's philosophical inquiry in his book *The Philosopher's Dog*, where he (1) advocates an original moral approach to animals, especially with regard to the concepts of honour and dignity in relation to animals, (2) places his inquiry in what he calls 'the realm of meaning' and (3) achieves these goals to a great extent through the use of narrative.

I will explicate what Gaita means by an inquiry in the realm of meaning, a praxis which he relates to Wittgenstein's idea that philosophical investigations are conceptual in nature. I will further argue that Gaita's manner of doing philosophy is not about 'conceptual' investigations in any standard sense of the term, but transcends the habitual distinction between empirical and conceptual inquiries, as this distinction operates in many post-Wittgensteinian discussions today.

In the penultimate section I will discuss the particular relevance of philosophizing in the realm of meaning in areas where our moral sensibilities are in a process of change. Last I will briefly discuss how and why narratives and narrative literature are natural companions for this kind of philosophical enquiry.

1. Dignity and *Disgrace*

In *The Philosopher's Dog* (and also in the biography *Romulus, My Father*) Gaita uses and develops a notion of dignity, and a perspective on animal life, which are conveyed in a narrative mode rather than by definition or argument.[1] He tells stories, mostly about animals and people he has known, and then reflects upon these stories in a broadly ethical vein. Only occasionally does he close his discussion around something that could be characterized as a philosophical conclusion. There is Orloff the dog, Jack the cockatoo, a herd of goats. And above all there is the father, Romulus Gaita, a practical moralist who naturally extends his compassion, his companionship and his notion of what dignity requires to his treatment of animals. Of course, dignity is not

---

[1] Raimond Gaita, *The Philosopher's Dog* (New York: Random House, 2005); *Romulus, My Father* (Melbourne: Text Publishing, 1998).

the only concept that comes into play in Gaita's discussions. But this, and the concept of honour, are probably the two which you will best remember after putting down the book, as they both describe a very intricate, and yet easily recognizable and intuitively appealing kind of respect in our dealings with animals, beyond what we believe about their inner lives or ability to think, feel and suffer.

The story which I will use to open my discussion, though, is not one of Gaita's own, but one which he has borrowed from J. M. Coetzee's novel *Disgrace*.[2] The central concept here is honour rather than dignity, but the aspect of this concept evoked is closer to dignity than, for example, the honour of a soldier (the glory of courage in combat) or, indeed, honour at the Oscar gala ('thank you, this is a great honour').

The novel *Disgrace* tells the story of David Lurie, middle aged, twice divorced, somewhat frustrated and overall a quite ordinary literary/intellectual communications professor, who after a rather pathetic affair with a student is expelled from his teaching post at a university. He leaves town for an indefinite period to visit his daughter Lucy in the countryside. She in her turn is struggling as a farmer, inexperienced and hard working, with her black neighbour as her occasional farmhand, and harbouring a substantial moral hangover from the country's colonial past and racist present.

The book is a multilayered and, like all of Coetzee's novels, highly philosophical piece, which describes the main character's struggle to face and come to terms with a variety of moral quandaries due to inequalities of power: the post-colonial situation in Lucy's relation to her neighbours; the tradition of male supremacy in Lurie's imbecile seduction of the reluctant student and in the gang rape of Lucy by a band of (black) burglars; the antidote of male supremacy in the quasi-feminist scapegoating of Lurie by his former employer; and finally, as one of Coetzee's favourite themes, the perpetual problem with the ways humans treat non-human animals.[3] This novel evokes a number of moral complexities, and with them a number of moral concepts, not the least of which are the dissonant pair of opposites suggested by the title: *dis*grace and grace. However, I will here focus on the agenda of honour and dignity set by Gaita in his use of the novel.

To be useful during his visit David Lurie takes a job at an animal shelter kept by Lucy's friend Bev: a kind, plain, even ugly, married middle-aged woman with whom Lurie reluctantly starts another affair. At the animal shelter one of Lurie's tasks includes aiding Bev in euthanizing unwanted dogs and driving the corpses to an incinerator. Lurie is not particularly animal friendly or generally squeamish, and the job is at first not unnecessarily hard for him. But soon he becomes captivated by morally ambivalent reactions to the task before him. Despite continuing to perceive the task as necessary, he finds himself trying 'to protect the dogs who have been killed from the dishonour that they suffer when their bodies are disposed of'.[4] When he drives a loaded combi with dead dogs to the incinerator,

---

[2] J. M. Coetzee, *Disgrace* (London: Penguin, 2000). Discussed by Gaita in *The Philosopher's Dog*, 91–7.

[3] For an interesting discussion of these themes in the novel in relation to Bernard Williams's *Shame and Necessity*, see Catherine Wilson, 'Disgrace: Bernard Williams and J. M. Coetzee,' in *Art and Ethical Criticism*, ed. Garry Hagberg (Oxford: Blackwell, 2008), 144–62.

[4] Gaita, *The Philosopher's Dog*, 91.

he himself loads them, one at a time, on to the feeder trolley, cranks the mechanism that hauls the trolley through the steel gate into the flames, pulls the lever to empty it of its contents, and cranks it back, while the workmen whose job this normally is stand by and watch.

On his first Monday he left it to them to do the incinerating. Rigor mortis had stiffened the corpses overnight. The dead legs caught in the bars of the trolley, and when the trolley came back from its trip to the furnace, the dog would as often as not come riding back too, blackened and grinning, smelling of singed fur, its plastic covering burnt away. After a while the workmen began to beat the bags with the backs of their shovels before loading them, to break the rigid limbs. It was then that he intervened and took over the job himself . . .[5]

The question is: why does he need to do this? 'For the sake of the dogs? But the dogs are dead; and what do dogs know about honour and dishonour anyway? / For himself, then. For his idea of the world, a world in which men do not beat corpses into a more convenient shape for processing.'[6] Lurie does not blame the workmen for treating the corpses disrespectfully, and neither does he take on guilt for exterminating the animals. But he is silently disturbed by the procedure, and equally, silently, disturbed by his own reactions. 'The dogs are brought to the clinic because they are unwanted: *because we are too menny*. That is where he enters their lives. He may not be their saviour, the one for whom they are not too many, but he is prepared to take care of them once they are unable, utterly unable to take care of themselves . . .'[7]

Dishonour? Indignity? The question is: are these concepts applicable to dead animals, or even to living animals for that matter? Confronted with these questions, as presented by Gaita and Coetzee, it may occur to the reader that these concepts are perhaps not common in contemporary moral philosophical discourse in general. Perhaps they have their allocated place in some applied ethics of care, or nursing ethics, when talking about preserving the dignity of people who are 'unable to take care of themselves'. But in central regions of modern normative ethics they are simply awkward. They are moral concepts of an old-fashioned kind, rich in meaning and assumption and connotation. But it is precisely to these regions of rich and complex moral meaning that Gaita and Coetzee want to take us by the questions concerning dishonour and indignity suffered by dead dogs. As Gaita puts it: 'The question that is at issue now is not how one should treat dogs, or whether one should avoid dishonouring them, but what it can *mean* to dishonour them, especially when they are dead.'[8]

2. The realm of meaning

By raising the question of how one can dishonour a dead dog Gaita and Coetzee lead us into an area of uncertainty which is both moral and conceptual in nature. Lurie's

---

[5] Coetzee, *Disgrace*, 144–5, cited by Gaita in *The Philosopher's Dog*, 92.
[6] Coetzee, *Disgrace*, 146.
[7] Ibid.
[8] Gaita, *The Philosopher's Dog*, 96.

dry self-mockery expresses the deep insufficiency of the socially given moral and conceptual resources for dealing with his tasks at the animal shelter. 'He saves the honour of corpses because there is no one else stupid enough to do it.'[9] His response is clearly moral, and yet what he feels compelled to do seems vain, superfluous, not required and not even obviously laudable.

The character Lurie's, the author Coetzee's and the philosopher Gaita's thinking in these borderline areas of settled moral meaning is designed to lead the reader to reconsider the form and direction of his own moral thought: not by argument and proof deducted from generally agreed-upon premises, but by means of a tentative search for meaningful ways to talk about the constantly unfolding complexities of our moral situation.

In his attention to this register of meaning Gaita exhibits a combination of his two major sources of philosophical influence: Ludwig Wittgenstein and Iris Murdoch. Although the direct references to Murdoch are scarce, the book could in its entirety be characterized as variations on the Murdochian themes: the richness of moral meanings in everyday life, the omnipresence of moral and evaluative aspects in human language and attitude, the 'fat relentless ego'[10] interfering with a clear-sighted contemplation of the world, and the loss of a rich dimension of language and meaning in modern moral thought. Yet this is not what I will focus on here. Gaita's approach is perhaps more obviously an appropriation of Wittgenstein's thought to moral philosophy. He emphasizes that the most pressing questions concerning our moral relations to animals are not going to be solved by a better knowledge of animal consciousness, a better understanding of the nervous systems of specific species, or anything of this kind. No external, objective fact about animals will solve the central problems concerning our moral relations to them.[11] (Indeed, this is not even necessarily a question that can be 'settled' in any stable, context-independent sense.) What is called for is a conceptual investigation.

To illustrate his approach by means of contrast Gaita devotes a lengthy discussion to Eugene Linden's and Jeffrey Masson's respective books on 'animal consciousness'.[12] Both of these authors tell stories about animals to show them in a certain light: as beings very much like ourselves. Although Gaita is sympathetic to Linden's and Masson's ethical concern for animals, he finds their philosophical methods, and the epistemic and metaphysical presuppositions that underlie those methods, deeply confused. When trying to criticize scepticism concerning animal experience and a behaviouristic perspective on animals, he claims, these two writers tell stories that aim at proving that scepticism and behaviourism are untenable. In Gaita's words, they 'pile

[9]   Coetzee, *Disgrace*, 146.
[10] Iris Murdoch, *The Sovereignty of Good Over Other Concepts* (London: Routledge, 2001 [1970]), 51.
[11] For a similar line of argument, see Cora Diamond's reading of Coetzee's *The Lives of Animals* in 'The Difficulty of Reality and the Difficulty of Philosophy,' in Cavell et al., *Philosophy and Animal Life* (New York: Columbia University Press, 2008).
[12] Gaita, *The Philosopher's Dog*, 111–15. See also Eugene Linden, *The Parrot's Lament and Other True Tales of Animal Intrigue, Intelligence and Ingenuity* (Boston: Dutton, 1999) and Jeffrey Masson and Susan McCarthy, *When Elephants Weep: The Emotional Lives of Animals* (New York: Dell, 1995).

anecdote upon anecdote, unrelentingly, desperately, with barely a pause for thought or reflection'.[13] But, according to Gaita, by trying to prove something in this manner they implicitly conform to the standards of objectivity and appropriate evidence endorsed by the sceptics and behaviourists. On the one hand the evidence they offer will necessarily be inferior to scientific evidence, which in this view of objectivity is evidence par excellence. On the other hand the attempt to prove something will leave them stuck with misleading standards of inquiry for this kind of question, which is a question of our relationship to animals, our common life with them and what it means to treat another living being in cruel and random ways.

What really would be required, according to Gaita, is a radical break with the standards of objectivity and argument that are taken for granted in the natural sciences, in most parts of contemporary philosophy, and implicitly by Linden and Masson as well.[14] We can compare this to our relations to other humans: we do not need proof against scepticism about other minds, or evidence that the other's pain is like mine, in order to know that we should treat another human being with respect.[15] And what respectful treatment requires is not decided according to objective (preferably scientific) knowledge about human consciousness or by appeal to heartening stories that confirm that humans indeed do have the kind of inner life that warrants moral behaviour towards them. (All this would indeed be quite mad.)

Empirical evidence to substantiate a moral or philosophical claim is, thus, not what Gaita is after in his stories. The anecdotes gathered by Gaita have a different purpose: not to back up argument for some specific kind of treatment of animals, but to investigate our living with animals, and the ways in which our moral responses and our language make sense in specific contexts and situations. The important questions are posed in what Gaita calls the realm of meaning, and can only properly be contemplated there, with a sensitive grasp of the complexities of our lives with other beings, and the language we use to describe them.[16]

'The realm of meaning' is the kind of expression that runs a risk of becoming a piece of cumbersome philosophical jargon, and it is thus of some importance to note that it is used by Gaita in a quite rough and casual manner to describe the regions of thought where we focus on meaning rather than fact; where attention is given to what words mean, what *we* mean by our words and to what things mean to us. One way of investigating such questions in relation to morals is by entering the messy realm of thick moral concepts, like the concept of dignity, where descriptive and evaluative aspects intermingle in an inseparable way.[17] An investigation into such concepts is essentially an investigation into the place that they have in our moral and evaluative lives.

---

[13]   Gaita, *The Philosopher's Dog*, 114–15.
[14]   Ibid.
[15]   This theme, concerning other people, is developed by Stanley Cavell in, for example, the early article 'Knowing and Acknowledging,' in *Must We Mean What We Say?* (Cambridge: Cambridge University Press, 2002 [1969]).
[16]   For Gaita's use of the phrase 'realm of meaning', see the chapter called 'The Realm of Meaning' in *The Philosopher's Dog*, 99–108.
[17]   The idea of 'thick concepts' was introduced by Bernard Williams in *Ethics and the Limits of Philosophy* (London: Fontana, 1985).

Another path goes through storytelling, but the stories need to be something more than anecdotes with a pre-given argumentative point. They need to be something to meditate on, something open-textured to invite a re-organization of one's philosophical preconceptions and indeed one's concepts. Both thick moral concepts and stories are most at home in the kind of thinking which is complex and layered; common and shared, but nonetheless personal and idiosyncratic. The eye in this kind of inquiry is not on possible philosophical generalizations or solutions or proofs, but on a non-technical elucidation of moral experience and moral language.

Asking for the appropriateness of the concept of dignity or the concept of honour in relation to animals (dead animals!) is, on this view, not a technical or semantic issue to be solved with reference to agreed-upon criteria, but an investigation into our ability to respond to animals as fellow creatures. The concepts of dignity and honour, as we know them, cover aspects of our responses to how dead bodies are treated and how people are treated when sick, old and dependent. They have connotations of mental and bodily integrity, autonomy and separateness, but also of mutuality and respect. They are concepts which helpfully place themselves in the very middle of our concern for how to be separate beings in relation to others with whom we live in mutual dependence throughout our lives. But they are also concepts which extend to people we will never know, as well as to both wild and domestic animals. We find it intelligible, even important, though perhaps puzzling, that David Lurie, at the incinerator, should worry about the dishonour of the standard treatment of the dead dogs.

You can compare, for example, the dignity of the senior officer who is too dignified to make his own coffee, with the dignity of the aging grandmother who refuses to accept help with her daily chores. Here we can see the close connection between dignity and self-image. But we do not, perhaps, think of animals as having a self-image: there we will need to follow other paths of meaning.

Our notion of dignity is likely to change through experiences of ailment or deprivation. Living in a crowded refugee camp or working with severely disabled people would most likely alter it. This concept – like many thick moral concepts – is not only thick in the sense that it covers both a normative and an evaluative aspect, but also in the sense that it is layered. The meaning of dignity is something that one learns through layer upon layer of experience, conversation, reading and imagining. It is to some extent overlapping with the concept of honour, but the relationship between these concepts is not a matter of definition. Meanings, in non-technical, natural language, are reasonably stable but not fixed.

When pondering over the applicability of concepts like honour and dignity to dogs one is probably not asking if dogs have that special something which makes them potential subjects of dignity or indignity. One is rather thinking (or *should* be thinking, in Gaita's view) about what kind of sense one can make of the idea of dishonouring a dog.

3. Gaita versus Anscombe

To get a better grasp of Gaita's discussion of the realm of meaning it is helpful to contrast his use of 'dignity' (and 'honour') with Elizabeth Anscombe's use of the concept of dignity in her paper 'The Dignity of the Human Being.'[18] In this paper Anscombe argues mainly that abortion is an offence against human dignity. She ties the notion of human dignity to a conception of human life, and specifically human procreation, which is both naturalistic and teleological, as it builds a normative structure around some facts about what human procreation is like when unaffected by human will and modern knowledge. On her account, any purposive alteration of the natural course of human sexuality and procreation – such as is involved in abortion, contraception, change of sex and in vitro fertilization – is an assault on human dignity because (respect of) human dignity is essentially dependent on our living the life for which we are destined by our biological constitution in natural circumstances. '[I]f we don't stick to human procreation of human beings, we generate further contempt for beginning human life and further alienation from belief in the dignity and value of humanness.'[19]

The argumentative movement here is one of narrowing down the range of potential meanings to one that is arguably the central or essential one: a meaning of 'human' and a meaning of 'dignity'. The trouble with this is that the advocated procedure, in a strange way, tends to render the discussed concept of dignity impotent. It is weighed down by Anscombe's very specific moral perspective (a combination of her Catholicism and neo-Aristotelian naturalism) and her specific teleology of human life. But our everyday, non-technical notion of dignity precedes and transcends such frameworks, and cannot be adequately illuminated by the use of a defining framework. Because of the locked normative positions that are reflected in the understanding of dignity in Anscombe's discussion, few people who do not share her background assumptions are likely to gain much positive insight from what in spite of her Wittgensteinian influence amounts to a definition of human dignity. The paper's genre is more that of a campaigning pamphlet than a philosophical article, as its mode of argument is rather more straightforwardly moral or political than philosophical or reflective, although she argues, at least partly, with an academic kind of sophistication. A pamphlet can be highly thought-provoking, and for some people Anscombe's paper surely is that. But although the paper gives pride of place to the concept of dignity, it reduces the concept to one that is specifically designed to play a part in Anscombe's normative teleology.

By attempting to attribute a specific, supposedly pre-given content to the idea of human dignity, Anscombe fails to open up the concept for someone who does not share her particular moral outlook. The rich, layered, thick character of this concept is obscured, and thereby much of its moral and philosophical potential is overlooked as well. This of course is not unique to Anscombe: an attempt to define the meaning of a moral concept for purposes of philosophical clarity is often likely to desiccate the moral and philosophical potential of the concept, and leave it substantially less helpful for our purposes of moral elucidation and understanding. Unfortunately this is what philosophers often end up doing to concepts which, in their natural environments (such as everyday talk, literature, journalism, and so on), are juicy and full of possibilities. The major argument for this kind of defining practice is that this is how we gain precision

[18] In G. E. M. Anscombe, *Human Life, Action and Ethics: Essays* (Exeter: Imprint Academic, 2005), 67–73.
[19] Ibid., 72.

in philosophical thought. The major argument on behalf of an approach like Gaita's could be that precision here is not a matter of fixing meaning beyond natural language; rather, it is a matter of better understanding what we ordinarily mean by certain words, the roles that they play for us, and how they connect with a variety of questions that arise in the course of our lives.

4. Openness in a post-Wittgensteinian tradition

Reflection on dignity and honour – for example, on what it means to treat or refuse to treat the bodies of dead animals in certain ways – does not in Gaita's account call for an attempt to give a systematized or action-guiding account of the proper place of these concepts in our lives and moral vocabularies. It is rather, as Gaita puts it, 'reflection on how we live our life with this part of language'.[20] It is open-ended.

I believe this is a fruitful path of inquiry and it is *very* tempting to go on from here to celebrate the openness of this manner of philosophizing and the post-Wittgensteinian tradition of which it is a part, as opposed to the argumentative closures which are such an essential feature in Anscombe's paper as well as much of contemporary Anglo-American moral philosophy. This is how some post-Wittgensteinian philosophers sometimes understand their own practice, 'open' as in 'open-minded'; a self-description which is found deeply disturbing by many who do not share a certain post-Wittgensteinian vocabulary and point of view.

But like any form of reflective writing, philosophical writing consists of openings and closures, and one mark of good writing, and good philosophy, is that these are sensitively distributed in a way that is responsive to both the subject matter and the intended audience. Analytic philosophy typically strives for a form of argumentative/theoretical closure which some philosophers influenced by Wittgenstein tend to find utterly unhelpful. In contrast to these, as well as in contrast to the great philosophical systems of modern history, philosophers like Gaita work in an open-ended manner. But one should emphasize that the idea of openness or open-endedness that is relevant when characterizing this style of philosophizing is *methodological*. It has to do with a specific conception of philosophy as work on open-ended, everyday, natural language.

Yet, the notion of openness easily brings along a moral charge: open as opposed to closed, narrow minded, dogmatic. What is really a description of a conception of language and philosophical method becomes thus, if carelessly used, a self-congratulatory chant. This, I believe, is not a serious problem in Gaita's own work, but it is a real hazard when one tries to describe and replicate Gaita's manner of philosophizing.

Gaita (like Coetzee and also Diamond) locates the question of our moral relations to animals in the conceptual field of dignity, respect, recognition and mutuality rather than, for example, reflections upon objective, scientifically discoverable features such as animal consciousness or ability to feel pain. Gaita's conceptual field deals with relating

[20] Gaita, *The Philosopher's Dog*, 97.

to the world as a human being in a world of other beings, rather than with accumulating facts about other beings and using these facts as a ground for general recommendations for the treatment of animals. This kind of conceptual inquiry requires that the variety of meaning in natural language, its internal inconsistencies and oddities, are preserved in the course of inquiry: we investigate our life by attending to our life in language. His way of framing the questions of morality requires a different way of proceeding with the inquiry than that which has been standard in much of modern Anglo-American moral philosophy. What is essential here is the redirection of thought from the *question–argument–proof–definition–answer* model of philosophizing, to a close attention to a variety of nuances in life and language.

The view I am attributing to Gaita can be seen as a version of what Wittgenstein claims in the much quoted, though puzzling, passage where he emphasizes that philosophy does not deal with anything hidden, that anything it can find is already open to view. 'Philosophy simply puts everything before us, and neither explains nor deduces anything. – Since everything lies open to view there is nothing to explain. For what is hidden, for example, is of no interest to us.'[21]

Philosophy in this tradition redirects our gaze towards what we mean and what we can mean by the words we use in our inquiry. And meaning, in a sense, is nothing hidden. But as we are changing creatures in changing circumstances, it is inevitable that the nature of our relations to each other and to animals and the meanings of the words and expressions we use to describe these relations, are not in any simple sense plain to our view. Redirecting attention from theories and explanations to meaning will not liberate us from puzzlement but sets us rather on a different course of exploration. This evokes another well-known contention of Wittgenstein's, that philosophy is work on oneself: 'As is frequently the case with work in architecture, work in philosophy is actually closer to working on oneself. On one's own understanding. On the way one sees things. (And on what one demands of them.)'[22]

In philosophy, we struggle, each with our own understandings, not in the sense that philosophy is private or ineffable, but in the much more everyday sense that we, ourselves, must try to understand what we have before us, and talk about it in words that we have made our own.

5. Beyond the conceptual and the empirical

But there is something about this picture of Gaita's post-Wittgensteinian perspective that is overly neat, and conforms too comfortably to a received use of Wittgenstein's philosophy in these kinds of discussions.

It is a common and sometimes warranted charge against philosophers inspired by Wittgenstein, or so-called ordinary language philosophers more widely, that they use

[21]　Ludwig Wittgenstein, *Philosophical Investigations*, 2nd edn (Oxford: Blackwell, 1958), §126.

[22]　Ludwig Wittgenstein, *The Big Typescript: TS 213*, ed. and trans. C. Grant Luckhardt and Maximillian A. E. Haue (Oxford: Blackwell, 2005), §86.

their own fallible intuitions as more or less infallible guides to actual language use. Gaita is generally free from this particular vice, but there is an interesting passage in *The Philosopher's Dog* where it seems to me that he does fall into this trap, and thus offers a misleading description of his own endeavour. He suggests that, in Linden's and Masson's respective books of animal anecdotes, they seem to be addressing the problem of our relations with animals in terms of evidence, as though we needed to know what animals are like. Although they 'start from a belief that the assumptions of behavioural science about objectivity and evidence actually distort our understanding of animal life, they are none the less in the grip of those same assumptions to the degree that their work remains answerable to the standards of evidence they believe it is wrongheaded to demand'.[23] Gaita, for his part, suggests that the problem is properly located in the realm of meaning, as a conceptual problem. But Linden and Masson do not see, he suggests, what kinds of absurdities their mode of inquiry leads them to. Masson supposes, for example, that spiders may well have a rich inner life (that this is something we can find out about spiders), without seeing that there is something odd with this very idea. On Gaita's view you could not properly attribute a rich inner life to a spider, not because it would be obviously untrue but because it is utterly unclear what it would mean.[24]

Thus, Gaita concludes, Linden and Masson must have overlooked, or be unable to grasp the proper distinction between empirical and conceptual questions. The idea of spiders having rich inner lives, for example, is not nonsense in the (factual) sense that it would be untrue; it is rather nonsense in the sense that it goes against meaningful language use. To illustrate this distinction Gaita produces another example. He suggests that his reader should consider two claims: (1) that one believes that there are unicorns and (2) that one believes that there are mice like Mickey Mouse ('who speak and fall in love, who squabble with their neighbours and who drive around in little cars'). Both, he suggests, are nonsense, but in two very different ways. The first one is of a kind that goes against established facts (nonsense in the sense of obviously untrue) while the second one is of a kind that goes against 'what one can intelligibly say, what one's words can intelligibly mean, given the lives those words have in our language' (and is thus nonsense in a conceptual sense).[25]

One initial problem here is that this example is ill chosen: few people without a philosophical training would be much helped by it, as the distinction itself, between empirical and conceptual questions, is substantially clearer than the purported illustrative example. People with philosophical training, meanwhile, are quite likely to find the example farfetched: it prompts one to think about the metaphysics of mice and unicorns instead of giving a clear and incontestable illustration of the distinction.

But the real, deeper problem is that the distinction between conceptual and empirical questions, as introduced here by Gaita (and a common device for distinguishing a

---

[23] Gaita, *The Philosopher's Dog*, 112.

[24] Ibid., 133–4. Masson writes, for example, that 'A mother wolf spider is as kind to strange baby wolf spiders as to her own. This might or might not be accompanied by an emotional state' (*When Elephants Weep*, 68).

[25] Gaita, *The Philosopher's Dog*, 133–4.

certain kind of philosophical investigation in the spirit of Wittgenstein's work) is ultimately unhelpful for illuminating Gaita's idea of philosophy as work in the realm of meaning. This is because much of the interesting work in the realm of meaning is work on language which is not fixed, not already there for us, but rather in a process of becoming.

Let us take a closer look at this. The habitual (analytic) distinction between empirical and conceptual questions ('Is the cat on the mat?' vs. 'Are bachelors unmarried?') pictures conceptual questions as something quite simple and clear: the answer is already there before our eyes, it is part of our ordinary competence as speakers. Yet when Gaita talks about conceptual questions he has a somewhat broader category of questions in mind. That stones or trees do not have inner lives is in this extended sense a conceptual issue: it follows from what it means to be a tree, from how the word 'tree' works in our language, that a tree does not have an inner life.

But even this extended notion of 'conceptual questions' is too simplistic to catch what is peculiar about David Lurie's pondering over the dead dogs. Lurie's work must be done in the realm of meaning: it is about what the things he does and the words he uses mean to him, and about the applicability of words. But it is not merely conceptual work in any straightforward sense, even if we invoke Gaita's extended notion of conceptual questions. Lurie is surely not in want of a definition: thus far, any post-Wittgensteinian philosopher would agree. But neither will his puzzlement be solved by attention to how we talk, because it is not just about how we actually do talk: it is also about how we (he) could (and should and will and want to) talk. The conceptual–empirical distinction is used by Gaita to discredit the introduction of empirical evidence when arguing for a specific way of viewing animals. Gaita suggests that we should see this as a place for conceptual work, of a type which is indifferent to empirical evidence. But in those unsettled regions of meaning where David Lurie (or Gaita himself, or Linden or Masson) makes his discoveries it is not at all obvious that empirical evidence is out of place.

It is not obviously nonsense to suppose that various animals, starkly different form us, have rich inner lives, since the very idea of an inner life, and the whole vocabulary and imagery that goes with it, is unsettled in these kinds of discussions. We have here an implicit, ongoing re-negotiation of applicability, where *both conceptual investigations and empirical evidence* may be relevant.

Empirical evidence (in a broad sense) has always changed the ways humans use words; it has changed our language, our concepts and our vision of things.[26] Thus empirical evidence should not be shut out from an investigation which purports to clarify how we should think and talk about our moral relations to animals. Even though these moral relations are more helpfully discussed in the realm of meaning, this does not mean that empirical evidence is beside the point.

When Gaita does philosophy in the realm of meaning, he is, indeed, most often moving beyond the habitual distinction between empirical and conceptual questions.

---

[26] A related point is made by Ian Hacking in 'Deflections,' which is his response to Cora Diamond's paper 'The Difficulty of Reality and the Difficulty of Philosophy,' in *Philosophy and Animal Life*.

He operates with a much broader idea of what counts as a conceptual question, than what is implied in the 'bachelors are unmarried' formula, but also, implicitly, with a broader one than the standard post-Wittgensteinian practice of surveying our various everyday ways of talking. I think this movement beyond the given formulas is a good and constructive one, yet, as I have been arguing, we (and Gaita) need to be careful not to retreat back into those narrow formulas when raising objections to those who wish to deploy empirical data in their philosophical arguments.

Linden and Masson may be unpersuasive. They may focus on the wrong things, and be insensitive to conceptual issues, but their mistake is not that they attempt to tell us something about how and what animals are. This is not a mistake, because what we believe about animals *is* relevant for how our relation to them develops over time, and thus relevant for which concepts are applicable in our relations to them. David Lurie's moral effort in *Disgrace* is not conceptual as opposed to empirical: it is essentially both. This is perhaps more obvious in his relations to other humans than in his relation to animals, but it applies equally to both. A new piece of information – such as that his favourite prostitute is also a mother, for example – opens up new paths of thinking for him. The successive re-organization of his concepts makes him susceptible to different kinds of information, which again enable him to alter his vision and his concepts further.

This is a very intricate process when it is served to us in the fast moving medium of a novel, but in fact we all go through these kinds of transformations in our lives, usually at the slow pace of years and decades.

6. Moral change

Paying attention to the mobility and suppleness of our concepts in the use of natural language is likely to strike us as particularly compelling in areas of life where our moral sensibilities and precepts are under reconsideration. In such areas any philosophical attempt to fix the content of specific moral concepts, in a way analogous to Anscombe's treatment of the concept of dignity, is likely to lack credibility: there will be more ambiguity, tentative approximations of meaning, conflicting intuitions and fewer people to applaud a simple normative/definitional closure around central moral concepts, unless the philosopher succeeds with some kind of conceptual *coup d'état*.

We may consider, with Gaita, a tentative case of change in our moral concerns. We may wonder today how well the concept of dignity is applicable to dead stray dogs. We easily apply the concept to at least some of our pet animals, and some particularly charismatic wild animals, like lions, though perhaps not without the feeling that this may be a secondary, derived way of speaking. Yet, as Gaita notes:

> One day – and it may not be too far away – we may look with revulsion at the cruelty of many of our present practices in relation to animals. But we may also become deeply ashamed of how impoverished our sense was of animal dignity. We

may become incredulous that we could ever have left animal corpses on the road to be run over again and again.[27]

If this were to happen, would we call this a conceptual change? Not necessarily. But it would be a change of our ways of talking about and relating to animals. Is this, then, conceptual change? Over time it would certainly amount to that; there would eventually be nothing (conceptually) inappropriate about applying the concept of dignity to animals. This kind of extension of what dignity requires looks quite conceivable from the point of view of our compassionate relations to some animals, but it is made to look like a conceptual impossibility if we try, in philosophy, to clarify our current practices through definition, by freezing a notion of dignity where animals can be included only in a secondary sense (with the kind of self-mocking uncertainty that David Lurie reveals when feeding his load of carcasses to the incinerator).

To make sense of moral change one must be sensitive to the mobility of the language we use to talk about moral issues: which words do we use – and when do we use them – for moral specification, or for praise or blame? How do our historically conditioned sensibilities translate into words and sentences? These considerations take us, finally, to consider the striking historical changes that make our relations to non-human animals today a complex area of moral and conceptual re-negotiation. I will here outline four aspects of this change, which I believe constitutes an important background for Gaita's inquiry.

First, reliance on animals for food and other utilities in relatively affluent late modern societies differs considerably from reliance on animals in earlier times. The industrialization of animal keeping has created a distance between people and the animals they eat. When animals (in the Western world, as recently as the 1950s) were brought up close to human households the moral duties and moral costs of meat, milk and skin were part of everyday life for a large part of the population. Children were involved in caring for animals, and slaughter implied the disappearance of cared-for pigs and calves. Meat was precious in more than one way. Today most people in the Western world get their animal protein in neat and affordable plastic packages from the supermarket – no strings attached.

Second, whereas people (especially in northern or otherwise less hospitable climates) used to be dependent on animals for sufficient amounts of crucial nutrients, contemporary means of food preservation and transportation have today made it possible to live well on a vegan diet anywhere, at least in relatively affluent parts of the world. Such people now have a real choice.

Third, a modern expansion of moral concern and attention to inequalities of power, which have rendered racism, class inequalities, sexual inequalities and child abuse deeply embarrassing for the late modern sensibility, are likely to exercise some influence on our relation to animals. Here we find the appeal of Peter Singer's idea of an expanding circle of moral concern:[28] The inner logic of modern liberal, egalitarian

[27] Gaita, *The Philosopher's Dog*, 37.
[28] Peter Singer, *The Expanding Circle: Ethics, Evolution, and Moral Progress* (Princeton, NJ: Princeton University Press, 2011 [1981]).

concern for the well-being of everyone puts us on guard against differentiating criteria. Why would animals not be part of 'everyone'?

Fourth, more and more attention is being paid to the ecological impact of animal protein in comparison to vegetarian food. Although this is not directly related to our relations to animals we can see today how ecological considerations add a moral burden to the practice of meat eating. This again pushes public opinion on this matter towards recognition of the issue of meat eating as morally problematic, which in its turn invites us to consider our current treatment of animals as problematic.

Our conception of animals and our moral relations to them are currently undergoing a re-negotiation. An increasing number of people in the industrialized or post-industrial world – including vegetarians and animal lovers, but also people who fall under neither of these descriptions – are beginning to think and talk differently about how animals are treated by humans. And in any such re-negotiation of moral concerns it is absolutely vital to pay attention to the mobility, suppleness, indefiniteness of the concepts we use to articulate our moral conceptions and insights.

7. Literature and the re-negotiation of moral conceptions

Now finally, we should note that it is not incidental that Gaita should be using stories and literature for his inquiry which (1) is pursued in the realm of meaning, and (2) investigates a region of morality where our sensibilities are in a process of change. It has been suggested that narrative literature (like narratives in general) is much better at putting words to moral quandaries in cases where our sensibilities are conflicted. As Martha Nussbaum puts it, for a certain category of perspectives on life: 'a literary narrative of a certain sort is the only type of text that can state them fully and fittingly, without contradiction'.[29]

Literature can approach morality from the inside of experience rather than from an abstracted exterior point of view of philosophy or science. The internal logic of narrative literature does not require that its key terms are defined: it can describe human behaviour without pressures of consistency. It utilizes the suppleness and mobility of our concepts; Coetzee or Gaita or Lurie (the author, the commentator and the character) can ponder over the suitability of a word like 'dignity' without needing to pin down its precise conditions of applicability. Thus it should be no surprise that literature and the narrative form have become central companions to a certain kind of philosophers in their inquiry into our relations to animals. But there are several ways of understanding this relationship, which are partly overlapping and complementary.

One habitual way of understanding the usefulness of the narrative form, developed by Nussbaum among others, is that it is animated; that it engages the emotions and can thus better move us to compassion.[30] Another way of understanding it is, as suggested

[29]   Martha Nussbaum, *Love's Knowledge* (Oxford: Oxford University Press, 1990), 7.
[30]   See, for example, Martha Nussbaum, *Upheavals of Thought: The Intelligence of Emotions* (Cambridge: Cambridge University Press, 2001).

by Cora Diamond, that literature can provide 'paradigms of a sort of attention';[31] it can show how we can attend to animals in a quite different way, see them differently than we would, had we not paid attention of the relevant sort. A third way of understanding the appeal of narratives, and of literature, when elucidating our relationship to animals, is this: in an area where our intuitions and sensibilities are blurred, conflicted or changing, we are sometimes likely to benefit more from the conceptual work of a literary artist than from that of a philosopher. In such cases the philosopher can often formulate ideas that match only some aspect of our intuitions; we are inclined to say 'yes, but . . .'. Literature explores our language, our uncertainties or inner conflicts. It 'holds the mirror up to nature'[32] and tentatively, suggestively, provides us with words and images to think with.[33]

But literature, of course, is not only a vehicle for the re-negotiation of moral conceptions, identities and world views over time, diachronically, but also constitutes a great reflective moral presence here and now, synchronically. The concepts of dignity and honour, at the heart of Gaita's inquiry, are as I said earlier, old fashioned, thick, resonant with multiple meanings. Lurie's reaction to the dead dogs is not the self-assured moral reaction of a young vegan raging against his conservative, meat-eating parents. It is tentative: he wants to say one thing, and yet doesn't know how what he wants to say can make sense in the face of other things that he thinks and says and knows.

Literature can formulate what is changing, but also what is resistant to change, what surprises us in its persistence. Our, and Gaita's, particular interest in Lurie's response is part and parcel of a modern sensibility, a growing preoccupation with our relations to and use of animals. Yet the need to relate to other beings is perennial and unavoidable. The rules and expected forms of compassion and co-creatureliness change over time, but the recognition of life in another being, and its death, the indignity and dignity and vulnerability and terribleness of a corpse, persist.

I would like here to challenge the imagery of dependence, oppression and mercy evoked by the dead dogs at the incinerator, and give an example where the proximity of animals and slaughter and danger and death in yesterday's rural society is combined with the sense of the dead or dying animal as something both great and terrible, awesome, *deinos*. Here is a cure for fears as described by the Swedish author Sara Lidman, in her novel *Naboth's Stone* (*Nabots sten*), to be used if a child would happen to develop an irrational habit of fear after it has (by the age of nine or so) outgrown the tales of supernatural creatures that grownups might scare him with to keep him away from dangerous places.

---

[31]   Cora Diamond, *The Realistic Spirit* (Cambridge, MA: MIT Press, 1991), 299.

[32]   This Shakespearean expression (from *Hamlet*, Act 3, scene 2, 17–24) is fondly used by Iris Murdoch in relation to art more generally, in an interview with Bryan Magee, published in Iris Murdoch, *Existentialists and Mystics: Writings in Philosophy and Literature* (London: Chatto & Windus, 1997), 12.

[33]   This is by no means unique today for the discussion of our relations to animals. Literature as a tool for moral re-negotiation in a situation of change is also the theme of Robert Pippin's book *Henry James and Modern Moral Life* (Cambridge: Cambridge University Press, 2000), where he explores Henry James's re-negotiation of the social order in the wake of the liberal, bourgeois subject.

There's a cure for that.

The cure is that the child shall take part in a slaughtering – or at least in the cutting-up. The slaughterer then points out a little muscle on the inside of the shoulder, close to the backbone, a special little muscle that keeps jerking long after the animal had been bled and gutted and flayed. The child has to creep inside the steaming carcass and bite that muscle, keeping hold of it between its teeth until it stops trembling

then

after that bite all imaginable fear of the dark vanishes

and the child dares everything, at any time . . .[34]

We no longer, generally, protect our children by frightening them, and we no longer think it appropriate to bully them out of their fears, even in theory, even as a joke. Our children are just as Lidman's rural hero Didrik wants his bride Anna-Stava to be: fed with animals they have never seen until the creatures are presented to them in thin slices at the table.[35] Quite like David Lurie, that is: foreign to the perspective where the human child, needy but intent on surviving, approaches the animal – and its dreadful but necessary death – with awe.

Lurie's awakening protectiveness of the dignity of the dogs can be seen as a raising of his consciousness about the moral tensions and the tensions of power inequality in his immediate surroundings; indeed, in his *life*. It can be seen as part of his personal pilgrimage. This is how I believe that it is usually read by philosophically minded readers. But as the novel vividly shows, his vaguely condescending pity will not suffice as a moral basis for his relations to the various human dogs of the novel: himself, Lucy, the farmhand, the robbers and rapists, the seduced girl. Maybe even his initial notion of honour in relation to the dogs is too meek and too aloof, too clueless to shake him out of his complacent superiority. Maybe Coetzee talks of honour rather than dignity (in a place where both would clearly be appropriate) to evoke the belligerent aspect of the former word: we see in our mind's eye the object of our aloof moral concern – college girl, black boy, stray dog, cow – fight back with her/his/its last shivering muscle, against oppression, but even more against the pity of the enlightened, compassionate oppressor.

Maybe.

Attention to these kinds of issues is philosophizing in the realm of meaning. This is the infinitely complex region of thought and vision and imagination, where we are constantly wrestling with the world and the limits of our own understanding, and which Gaita generously invites us to enter.[36]

---

[34] Sara Lidman, *Naboth's Stone* (Nabots sten), trans. Joan Tate (Norwich: Norvik Press, 1989), 30–1.
[35] See Sara Lidman, *Vredens barn* (Stockholm: Bonniers, 1979), 12.
[36] I am thankful to Niklas Forsberg and Mikel Burley for thorough comments on earlier drafts of this essay.

13

W. G. Sebald and the Ethics of Narrative

Alice Crary

W. G. Sebald is well known for his preoccupation, in his prose fiction, with individuals who were ripped out of the circumstances in which they were at home by wars, colonial enterprises and other cataclysms of the twentieth century. A generation of readers has drawn attention to Sebald's engagement with ways in which people cut off from their pasts by different forms of political upheaval are made to suffer. Many readers have also underlined Sebald's distinctive narrative style, bringing out how, among other things, he uses narrators whose biographies match his own in specific respects, scatters photographs throughout the resulting narratives, adorns these narratives with digressions that explore bizarre and suggestive coincidences and avoids the use of direct quotation. Those who describe the characteristic moral themes and literary strategies of Sebald's prose narratives often do so with an eye to demonstrating that the narratives have substantial moral interest. In this essay, I defend a version of this image of Sebald as a morally consequential author. My guiding thought is that, if we are to account for the ambitions of Sebald's fictional oeuvre, we need to do more than track the development of its central themes and literary strategies. We need also to capture the manner in which these thematic and literary elements complement each other. Sebald's works of prose fiction have unusual narrative structures. They are designed to engage readers in complex ways, and their strategies of engagement contribute in non-accidental ways to what they reveal about the harms on which they focus, harms inflicted by political forces that tear people out of the settings in which they first encounter the world.

This claim about Sebald's prose narratives is equivalent to a proposal to take certain literary works to contribute internally to genuine or rational understanding specifically as works of literature (i.e. specifically as works that tend to engage readers emotionally in various ways). While proposals along these lines are not unheard of in conversations about philosophy and literature, they encounter deep resistance both in philosophical and in critical circles. At the most fundamental level, this is because they challenge a deep-seated understanding of the notion of rationality. Philosophers and critics generally at least tacitly assume that a bit of discourse qualified to directly inform rational understanding must do so in a manner independent of any tendency it has to engage us emotionally and, by the same token, that it must be possible to follow a rational line of thought in a manner that doesn't essentially depend on the possession of any particular capacities of emotional response. What counts as rational, within the context of this assumption, is *argument*, where an argument is taken to be a statement or set of statements from which a further concluding statement can be inferred in a manner that doesn't essentially depend on any tendency of the initial statement or statements to engage us emotionally. Not that this way of talking about argument is obligatory. Some thinkers speak of argument more broadly in reference to aspects of literary and philosophical works that, as they see it, contribute internally to our ability to make particular inferences by directing our feelings.[1] But it is argument in the narrower sense just described that is internal to the conception of rationality that seems to speak against allowing literature as such to directly inform genuine understanding. For the sake of convenience, I here speak of 'argument' in connection with argument in this relatively narrow sense. I refer to the conception of rationality that limits rationality to argument thus construed as the *narrower conception* of rationality, and I refer to the sort of contrasting conception that allows rationality to include more than such argument as the *wider conception*.[2]

The narrower conception of rationality is what at bottom drives resistance to the idea that a literary work, such as one of Sebald's prose narratives, might directly inform moral understanding in virtue of ways in which it engages us. There can be no question of making room for this idea apart from availing ourselves of the resources of a more permissive conception of rationality. Philosophers and critics who represent works of literature as capable of directly informing understanding qua works of literature at least implicitly draw on the relatively permissive conception of rationality that I am describing as 'wider'. Some of these thinkers explicitly draw attention to their procedures and present arguments in favour of this conception.[3] I believe that a successful argument for the wider conception can be produced, but I do not supply one in this essay. This means that I do not directly address concerns of those philosophers and critics who accept the constraints of the narrower conception of rationality and who believe that, even if we can show that a given literary work is naturally read as having intelligent powers that are direct functions of its tendency to engage us, we cannot show that we are justified in crediting it with such powers. I regard as cramped and indefensible views that

---

[1] See, for example, Martha Nussbaum's discussion of what she calls the 'arguments' of some of Henry James' novels in 'Introduction: Form and Content, Philosophy and Literature,' in *Love's Knowledge: Essays on Philosophy and Literature* (Oxford: Oxford University Press, 1990), 3–53. Or see Simon Glendinning's remarks about the kinds of 'arguments' that, as he sees it, pervade classic phenomenology in *In the Name of Phenomenology* (London: Routledge, 2007), esp. 20–1.

[2] What in this paragraph I described as a narrower understanding of argument is capacious enough to encompass not only *formal* inferences but also what are sometimes called *material* inferences – that is, inferences that turn for their validity on the content of their premises and conclusions (e.g. 'The baby is very tired; she'll soon be asleep'). So if we conceive of rationality in terms of narrower argument we do not thereby commit ourselves to any sort of formalism. We simply claim that any stretch of discourse presenting a rational form of instruction must, without regard to whether the inferences it licenses are taken to be formally or materially valid, be recognizable as such apart from any emotional engagement with it.

[3] I discuss the relevant body of work in 'Literature and Ethics,' in Hugh Lafollette (ed.), *The International Encyclopedia of Ethics*, available online.

thus exclude on a priori grounds the possibility that a literary work might internally contribute to rational understanding as a literary work.[4] But in this essay I set aside the task of assembling philosophical considerations in favour of accommodating this possibility. I limit myself to showing that, if we are willing to at least allow that a work of literature might as such directly inform rational understanding, a good case can be made for thinking that Sebald's fiction does so. Sebald's prose narratives are plausibly read, in a manner that presupposes the wider conception of rationality, as designed to engage readers in ways that contribute internally to genuine understanding.

One thing that speaks for this approach to Sebald is the fact that his narratives presuppose the logic of the wider conception of rationality at the level of their narrative themes. A recurring theme of the narratives is that people who have been forcibly uprooted run into obstacles in cultivating their own sensibilities. As a result of having been violently disconnected from their pasts, Sebald's characters find themselves responding to their environments in ways that they cannot understand well enough either to decisively identify with or to repudiate them. A second and closely related theme of the different narratives is that the trouble these characters have developing their own modes of responsiveness interferes with their ability to understand their lives. The kind of self-understanding they are cut off from doesn't consist in and can't be reduced to the sort of neutral knowledge distinctive of the sciences. Participation in scientific and other practices of fact-gathering presents no special obstacles to them; the characters in question include, among others, a surgeon and a schoolteacher who is a serious amateur botanist.[5] Yet these characters lack the sort of insight into themselves that would allow them to truly live. Within the narratives, it is suggested that this lack of insight is an immediate expression of their emotional handicaps, and the point here is that this suggestion depends for its coherence on the logic of the wider conception of rationality.

The portions of Sebald's prose narratives that thus employ the logic of the wider conception of rationality present us with accounts of human life that are *naturalistic* in that they depict human beings as natural, non-magical beings or, alternately, as animals of a distinctive kind. It is widely recognized that naturalistic modes of thought play a significant role in Sebald's thought. But if we are to appreciate what is at stake when, at specific junctures in his writing, descriptions of human beings and other animals are presented as studies in naturalism, we need to recognize that the narratives

---

[4] Although philosophers and critics who advocate views of literature that make this exclusion invariably rely on the restrictions of the narrower conception of rationality, most do so only tacitly, without acknowledging that questions can intelligibly be raised about whether the narrower conception does justice to what rationality is like. This includes the authors of some of the best-received critiques of the idea that literary works can as such directly inform moral understanding. See, for example, D. D. Raphael, 'Can Literature Be Moral Philosophy?' *New Literary History*, 15 (1983), 1–12; Onora O'Neill, 'The Power of Example,' in *Constructions of Reason: Explorations of Kant's Practical Philosophy* (Cambridge: Cambridge University Press, 1989), 165–86; Richard A. Posner, 'Against Ethical Criticism,' *Philosophy and Literature*, 21 (1997), 1–27; and Joshua Landy, 'A Nation of Madame Bovarys: On the Possibility and Desirability of Moral Improvement Through Fiction,' in Garry L. Hagberg (ed.), *Art and Ethical Criticism* (London: Blackwell, 2008), 63–94.

[5] The surgeon and the amateur botanist are Dr Henry Selwyn and Paul Bereyter of Sebald's *The Emigrants* (New York: New Directions, 1997).

explore two fundamentally different types of naturalistic modes of thought. Some of these modes of thought belong to *natural history*. 'Natural history' is the name for the part of biology dedicated to sorting living organisms into kinds and describing their characteristic features and operations, and Sebald is clearly interested in natural history in this traditional sense. Not only are his narratives filled with characters who are trained in relevant practices of naturalistic classification, his narrators occasionally offer their own amateur natural histories of different organisms.[6]

But Sebald's narratives also deal in naturalistic modes of thought that don't belong to traditional natural history. The narratives invite us to take an interest in animate life in ways that, while resembling natural-historical thinking in being naturalistic, are also ethical in that they are essentially expressive of ethical value. For the purposes of this essay, I speak in this connection of *ethically naturalistic* modes of thought. Admittedly, the idea of such modes of thought is philosophically controversial. There is a general consensus among philosophers that the natural world is the world that is there 'anyway' – that is, apart from any direct reference to our sensitivities or modes of appreciation. Within the context of this philosophically influential understanding of the natural world, it appears that any mode of thought with a necessary tie to sensibility has an essential tendency to distort our view of how things are. If we assume that ethical thought is necessarily shaped by sensibility, then it appears that there is no room for undistorted modes of thought that are both essentially ethical and essentially naturalistic. But the philosophical understanding of the natural world that may seem to cause trouble for the very idea of ethical naturalism is not beyond criticism. If we developed a successful argument for what I am calling the wider conception of rationality, and if we thereby demonstrated that there is no such thing as making the connections internal to some undistorted lines of thought about the (genuine, natural) world apart from the possession of certain routes of feeling, we would at the same time have taken an important step towards discrediting this understanding of the natural world. I am not, however, here interested in defending Sebald's preoccupation with ethically naturalistic modes of thought. I want only to note that modes of thought of this sort are at play in passages in his writing in which depictions of individual human beings as possessing 'wider' capacities of reason are presented in a naturalistic light. In these passages, we are invited to look upon an individual's ability to develop such capacities as a good. For instance, we are encouraged to regard individual characters as having suffered tragic deprivations insofar as they have been subjected to violent forces that interfere with their growth in relevant respects. Nor is there any suggestion that we arrive at this view of things simply by projecting our attitudes or adopting perspectives that are merely accidently related to understanding.

---

[6] The naturalists in the narratives include Paul Bereyter of *The Emigrants*, who was one of Sebald's actual childhood teachers, the seventeenth-century naturalist Thomas Browne, who figures prominently in *Rings of Saturn* (London: Vintage, 1998) and the various members of the fictional Fitzgerald clan of *Austerlitz* (New York: Modern Library, 2001) who are botanists and naturalists of other kinds. Additionally, a significant portion of Sebald's prose poem *After Nature* (New York: Random House, 2002) is devoted to the life of the eighteenth-century botanist, zoologist and physician Georg Wilhelm Steller. Finally, the narrator of *Rings of Saturn* relies on the work of unnamed 'natural historians' in talking about the lives of herring and silkworms (see 49ff. and 274ff.).

To the extent that Sebald's narratives represent emotional development as internal to the cognitive growth of their characters, they encourage us to regard our willingness to explore the attitudes they themselves seek to cultivate as internal to activities of thought about how things really are.

Sebald's fiction abounds with ethically naturalistic images that take artefacts as proxies for human beings. His narratives frequently depict household objects, clothing and other personal effects and, above all, buildings as emblems of human action. In what qualifies as an ethically naturalistic flourish, the narratives frequently invite us to see these objects' vulnerability to displacement and decay as figures for the vulnerability and transience of human life. Strikingly, Sebald occasionally describes such ethically naturalistic treatments of artefacts as constituting a sort of 'natural history' of them.[7] This way of talking underlines the importance of the role of historical perspectives in ethical thought as Sebald conceives it. It is not a sign of any confusion on Sebald's part about the distinctness of the modes of thought proper to traditional natural history.

Human beings (and artefacts that stand in for them) are not the only things, in Sebald's fiction, that fall under an ethically naturalistic gaze. Non-human animals frequently do so as well. We are repeatedly called on to look upon animals of different kinds as meriting pity insofar as they are prevented from developing their characteristic capacities and truly living.[8] Moreover, it is a characteristic gesture of Sebald's to bring naturalistic and ethical images of animals together with naturalistic and ethical images of human beings. His narratives attempt to get us to recognize pathos in the lives of individual humans whose development bears scars of social upheaval by juxtaposing depictions of these individuals with descriptions of individual animals who have likewise been prevented from fully living.[9] While Sebald's narratives thus employ

---

[7] See in this connection the English translation of Sebald's 1997 lectures *Literatur und Luftkrieg*, which bear the title *On the Natural History of Destruction* (trans. Anthea Bell (New York: Modern Library, 2003)). In these lectures, Sebald describes the firebombing of German cities by Allied forces at the end of World War II. The lectures ignited controversy in Germany because Sebald was taken, wrongly, to be trying to mitigate atrocities of the Hitler government by pointing out that Germans were also made to suffer horribly during the war. In fact, Sebald set out to underline what he saw as avoidance of the horrors of the firebombing by German authors with an eye to bringing out ways in which this avoidance was connected to a failure to come to terms with the horrors of National Socialism. But I am not here concerned to resolve this dispute about how to read Sebald's lectures. The point I want to make has to do with the fact that, while within the lectures categories of ethical naturalism (and not those of traditional natural history) predominate, the English translation – which Sebald approved before his death – nevertheless includes the phrase 'natural history'. (For a comment about Sebald's acceptance of the English translation, see the 'Translator's Note' on p. x.)

[8] Sebald's narratives often move seamlessly between natural histories of animals of specific kinds and ethically naturalistic treatments of them. Thus, for example, in what he describes as a 'natural history' of herring, the narrator of *Rings of Saturn* integrates reflections about things 'natural historians' have said about herrings with evocative, ethically loaded observations about them (see 49f., esp. 56–7). The plausible suggestion of this passage, as well as of others like it, is that there is a connection between our ability to bring living things clearly into view using natural historical categories and our ability to appreciate the pathos of their lives.

[9] Thus, for example, in *Rings of Saturn*, the narrator gives us a portrait of members of the aristocratic Ashbury family who were impoverished in the financial crisis following the Irish Civil War and who, when the narrator encounters them, are effectively squatting in their ancestral home busying themselves with work that has about it 'something aimless and meaningless and [seems] not so

ethically loaded analogies between human beings and animals, they in no way undercut the idea that human life and human suffering are distinctive. On the contrary, the central, organizing preoccupation of the narratives is evoking special vulnerabilities we human beings have as reasoning animals whose lives are structured in distinctive ways by reflection and memory.

This ethically weighted, naturalistic vision of human life belongs at the centre of any adequate account of the moral and political interest of Sebald's thought. Central to the vision are the following two ideas, namely, the idea that human beings need to develop emotionally in order to think and live well and the idea that the destructive tendencies of political forces that uproot individuals from their surroundings are augmented by ways in which such forces interfere with emotional development. These ideas receive an especially subtle treatment in *Austerlitz*, Sebald's last prose narrative,[10] and in the remainder of this essay I am concerned with this work. My emphasis is on bringing out how the contribution *Austerlitz* makes to our understanding of certain specific human vulnerabilities is inseparable from an involved narrative strategy and how this book of Sebald's is aptly described as contributing directly to moral understanding specifically as a work of literature.

The focus of *Austerlitz* is the life and experience of a man, Jacques Austerlitz, who – like many of Sebald's characters – was displaced by World War II.[11] As the narrative unfolds, we learn the following about Austerlitz. In the summer of 1939, when he was four and a half, he was brought on a *Kindertransport* from Prague to London, where

much a part of a daily routine as an expression of a deeply engrained distress' (211). As readers, we are prepared to appreciate the pathos of this portrait of the Ashburys by an account, earlier in the narrative, of an isolated, caged Chinese quail 'running to and fro along the edge of [its] cage and shaking its head every time it was about to turn, as if it could not comprehend how it had got into this hopeless fix' (ibid., 36). (For a couple of additional illustrations of how Sebald's narratives use ethically rich descriptions of animal suffering to evoke the moral interest of certain forms of human distress, see my discussion of *Austerlitz* below.) Interestingly it is a presupposition of this Sebaldian strategy for leading us to see the pathos in human life (i.e. the strategy of moving from images of animal suffering to images of human suffering) that we respond more readily to animal than to human suffering. Given how cruelly and callously modern societies treat animals – think of factory farms and laboratories that do animal testing – this presupposition may initially seem implausible. Nevertheless, there is clearly something to be said for it. See in this connection, for example, Raimond Gaita's discussion of how passersby who witnessed first an elderly woman get struck by a car and then a dog get injured by the fall of a drunk man responded in greater numbers and with less restraint to the plight of the dog (*The Philosopher's Dog* (Melbourne: Text Publishing, 2002), 21–3).

10 The idea that *Austerlitz* is, in the respects just indicated, Sebald's masterpiece is not uncontroversial. I return to the topic of the place of *Austerlitz* within Sebald's literary opus in note 39, below.

11 Austerlitz, like most of the other characters in Sebald's different prose narratives, is partly fictional and partly drawn from life. The most significant real model for Austerlitz's life is the life of Susi Bechhöfer as recounted in Sally George's 1991 BBC documentary *Whatever Happened to Susi*. In interviews Sebald mentions the Bechhöfer documentary, and, while his memory of several of its details is inaccurate, he rightly represents it as providing a significant model for Austerlitz's years with his foster-family as well as for Austerlitz's later efforts to learn about his early life. See 'A Conversation with W. G. Sebald,' an interview with Joseph Cuomo, reprinted in Lynne Sharon Schwart, *The Emergence of Memory: Conversations with W. G. Sebald* (New York: Seven Stories Press, 2007), 93–118, esp. 110–11; see also 'The Last Word,' an interview with Maya Jaggi (*The Guardian*, 21 December 2001). Susi Bechhöfer wrote a memoir with Jeremy Josephs – *Rosa's Child: The True Story of One Woman's Search for a Lost Mother and a Vanished Past* (London: I.B. Tauris, 1996) – that contains a longer account of her life. But Sebald seems not to have read it.

he was taken to Wales to live with a childless couple, a Calvinist minister and his wife, who changed his name and told him nothing about his origins. By the time Austerlitz had finished at school, he knew his real name and had a few bits of information that he could have used to arrive at general facts about his background. But he repressed thoughts about where he came from both then and during his university studies and a 30-year long career as a lecturer on architectural history. He focused on his research, allowing the accumulation of knowledge to serve as a sort of 'compensatory memory',[12] until upon retirement it struck him that he had never understood his life and had 'never really been alive'.[13] Gripped by anxiety and thoughts of suicide, he began – at first without conscious exertion and later quite deliberately – to recollect and research his early childhood. He traced his family to Prague, managed to meet his one-time babysitter and uncovered facts both about how his Jewish mother was interred by the Nazis, deported and murdered and about how his father, also a Jew, initially avoided interment by travelling to France. When the narrative closes, Austerlitz is trying to learn more about his father's fate and to find a woman, Marie de Verneuil, who had been his own closest companion but to whom he never properly responded as a lover. There is no suggestion in the book that a person whose life has been fundamentally altered by murder and social upheaval, as Austerlitz's life has been, can heal and be whole. But it is suggested that the kinds of horrors in question do damage in part by interfering with the role memory plays in composing a coherent and fulfilling life and that a project of recollection can contribute to a significant, albeit necessarily partial, type of restoration. While Austerlitz's discoveries about his past don't affect any sort of happy transformation, some of them strike him with the force of small epiphanies, equipping him to make new sense of things and giving him an idea of what it might be to truly live.

Within the narrative, it is indicated that Austerlitz's break from his past interferes with his ability to cultivate his own sensibility and thereby obstructs his ability to make sense of his life. Again and again, Austerlitz's fractured personal history is depicted as putting him in a position in which he is subject to emotional responses that he cannot make sense of well enough either to reject or to claim them with conviction and in which, as a result, he is prevented from identifying with particular ways of responding to the world. Thus, for example, a dam near Austerlitz's Welsh foster-home elicits from him a jumble of conflicting and emotionally intense associations that he regards as an impenetrable mystery and that he only begins to relate to very late in life when, upon travelling up through Germany by the same route he took while being transported to England as a four-year-old, it dawns on him that he has always associated the dam with a tower he saw on his earlier journey.[14] Or, to mention a further example, a trip Austerlitz takes to the spa at Marienbad in 1972 with Marie de Verneuil gives rise in him to desperate feelings that he does not understand and cannot negotiate with conviction until he discovers that he and his parents visited the spa at Marienbad in 1938 and

[12]　Sebald, *Austerlitz*, 140; see also 260.
[13]　The inset quote is from ibid., 137; see also 44 and 125.
[14]　Ibid., 54–5 and 225.

realizes that his violent reaction is connected with associations from that earlier trip.[15] In these and other parts of Sebald's book, Austerlitz's rupture from his past is depicted as handicapping him in his efforts to cultivate his own routes of feeling, and it is also consistently suggested that it is in significant part in this way – by preventing him from arriving at a well-cultivated sensibility – that his dramatic loss of his childhood impedes his ability to understand his life and chart a course through it.[16]

The passages I just discussed describe Austerlitz as hindered in developing the sort of sensibility he would need for a good self-understanding, and I want to pause here to say a word about a number of other complementary passages. I have in mind a set of passages containing references to the later philosophy of Wittgenstein. Sebald's book has a narrator, a man about ten years younger than Austerlitz, and early on this narrator prepares us for the introduction of Wittgensteinian themes by including a photograph of Wittgenstein's eyes and commenting at length on what he sees as Austerlitz's 'personal similarity to Ludwig Wittgenstein'.[17] Later the narrator represents Austerlitz as talking about his own thought and experience using different images and catchphrases from Wittgenstein's philosophy. For instance, he describes Austerlitz as making use of the well-known image from Wittgenstein's *Philosophical Investigations* of 'rails laid out to infinity'. The narrator tells us that, in his efforts to uncover his past, Austerlitz goes to Theresienstadt, the camp, Austerlitz has learned, where his mother had been interred. On his way, he stands at a train station and is struck by the thought that the 'railway lines ran away into infinity'.[18] Further, in yet another nod to Wittgenstein's *Investigations*, the narrator represents Austerlitz as using the notion of 'family resemblances'. The narrator claims that Austerlitz describes his own method

---

[15]   Sebald's choice of the Marienbad Spa as the site of Austerlitz's different trips is evidently partly a nod to Alan Resnais' enigmatic 1961 film *L'année Dernière à Marienbad*. Resnais' film is a forerunner of *Austerlitz* in that, like Sebald's narrative, it deals with ambiguities surrounding a (possibly) forgotten past trip to Marienbad in connection with a romantic relationship. (For a helpful discussion of Sebald's interest in Resnais and in film more generally, see Matthias Frey, 'Theorizing Cinema in Sebald and Sebald with Cinema,' in *Searching for Sebald: Photography After W. G. Sebald*, ed. Lise Patt (Los Angeles, CA: Institute of Cultural Inquiry, 2007), 226–41. Sebald's selection of the Marienbad Spa seems also to have been prompted by the fact, noted in the text (see 210), that the waters at the spa bear the name 'Auschowitz Springs', a name not only close to 'Austerlitz' but also, hauntingly, to 'Auschwitz'.

[16]   There is no suggestion in Sebald's narrative that Austerlitz's growth is simply a function of his acquisition of certain plain facts about his childhood that he previously lacked (e.g. the fact that he once went to Marienbad with his parents or the fact that he travelled by train to England, over a certain route, on a *Kindertransport*). What the text suggests is that Austerlitz's growth is in significant part a function of the way in which, in the process of piecing together bits of his past, he learns to make sense of his own responses to the world and hence becomes capable of laying claim to and cultivating his own sensibility.

[17]   Ibid., 4 and 40–1. In these last two pages, we receive a list of resemblances that the narrator sees between Austerlitz and Wittgenstein. Like Wittgenstein, Austerlitz characteristically wears a 'horror-struck' expression on his face; carries a rucksack with him everywhere; is 'a man locked into the glaring clarity of his logical thinking as inextricably as into his confused emotions'; does with as few possessions as possible; and doesn't 'linger over any kind of preliminaries'. One thing that is not mentioned here but that we are justified in assuming is important for the narrator is the fact that, like the fictional character, the philosopher was violently uprooted by World War II.

[18]   Sebald, *Austerlitz*, 185.

in architectural history as one of finding 'family likeness' between buildings[19] and that Austerlitz frequently spends hours with a set of black and white snapshots he has taken and collected throughout his life and sorts them according to their 'family resemblances'.[20] Finally, in a gesture that is evidently a reference to the many passages in Wittgenstein's later writings concerned with the teaching and learning of simple mathematical series, the narrator relates how, when Austerlitz found his long lost Czech babysitter and heard her start to count in Czech, he realized that he himself could go on.[21]

To appreciate the interest of these passages from Sebald's narrative it is helpful to consider the portions of Wittgenstein's thought to which they refer. It is in an early passage of the *Investigations* in which he is talking about what it is to grasp the concept 'game' that Wittgenstein introduces the notion of family resemblances, observing that our efforts to grasp the concept 'game' invariably express our sense of the importance of resemblances between various of its applications.[22] To speak in this way of family resemblances is to say that a person's mastery of the concept 'game' is inseparable from her possession of an appreciation of the significance of similarities linking its different uses. One of the hallmarks of Wittgenstein's later philosophy is insistence on bringing the idea that underlies this talk of family resemblances in reference to games to bear on even those of our conceptual practices to which it might seem to be most foreign. In the famous set of remarks in the *Philosophical Investigations* that get referred to as the 'rule-following sections', Wittgenstein applies the idea to practices of counting or extending simple mathematical series.[23] It is in this context, specifically with an eye to evoking the class of views of mathematics he is challenging, that Wittgenstein speaks of 'rails laid out to infinity'.[24] His point is that, according to the views he repudiates, extending a mathematical series is like hitching oneself to a car on an indefinitely long mechanical rail and having it simply pull us along. We should instead see our ability to take successive steps as essentially informed by our sense of what speaks for them.

There is controversy about the upshot of these portions of Wittgenstein's thought. Many readers of Wittgenstein regard his willingness to represent sensitivities as internal to all our conceptual capacities as a sign that he doesn't think our conceptual capacities are matters of concern with genuine, objective regularities in the world. He

---

[19]   Ibid., 33.
[20]   Ibid., 119.
[21]   Ibid., 160.
[22]   Ludwig Wittgenstein, *Philosophical Investigations*, trans. G. E. M. Anscombe (New York: Macmillan, 1958), §67.
[23]   Why does the idea that we necessarily rely on sensitivities in picking out the extension of a concept seem to many philosophers to be excluded in cases of simple mathematical concepts? In many circumstances in which we apply concepts – say, the concept 'game' – we need to demonstrate a dual capacity, appreciating both the situation that confronts us and seeing what about it does or does not demand the use of the concept. Mathematical series seem less problematic insofar as they reduce to a minimum the first of these two elements. Although we need to locate ourselves within the series, there is no perceptual component to complicate matters. Whereas the idea that forms of training that instill responsiveness are internal to teaching might seem unobjectionable when it comes to the teaching of a concept like 'game', it can accordingly seem out of place with regard to mathematical concepts.
[24]   Wittgenstein, *Philosophical Investigations*, §218.

is here supposed to regard questions about the correctness of particular conceptual moves as questions not about fidelity to how things are but rather about something like community agreement.[25] Yet a fair number of other readers take Wittgenstein to be trying to get us to see that the relevant sort of reliance on sensitivities is fully consistent with genuine, objectively authoritative concepts.[26] Happily, we do not need to settle this exegetical quarrel in order to see which of the two interpretations at issue in it is operative in Sebald's book. The aspects of Wittgenstein's thought that surface in Sebald's book get mentioned in characterizations of Austerlitz. The book's descriptions of Austerlitz presuppose that an individual needs a developed sensibility in order to arrive at a good self-understanding. There is no suggestion in the text that the kind of self-understanding at issue somehow has community agreement as its touchstone or that it is anything other than genuine self-understanding. A person who held that no capacities that necessarily involve sensitivities can as such be matters of concern with the genuine, objective world could not consistently offer this portrait of Austerlitz. So there is every reason to think that, when bits of Wittgenstein's writing crop up in characterizations of Austerlitz, we should interpret them as invitations to regard Austerlitz's thought as necessarily involving his sensibility and as doing so in a manner that doesn't essentially compromise its status as a mode of engagement with what his life is really like.

That Austerlitz traffics in family resemblances in thinking about the history of architecture and in organizing his collection of snapshots is a narrative suggestion to the effect that he cannot help but draw on his own sense of salience and significance, however unstable and fragmented this sense is, in trying to make sense of his circumstances. One thing that may seem to speak against this suggestion is the thought that there are lines of thought that we can follow independently of the possession of any particular sensitivities. So it is noteworthy that the text hints at a challenge to this thought about transcendence of the human when Austerlitz is represented as making use of Wittgenstein's image of infinitely long rails. Insofar as the particular tracks Austerlitz has in view run to Theresienstadt, his gesture invites us to associate the illusion with the railway system at the heart of the Nazi machinery of death and hence with the very forces that displaced and damaged him, indelibly marking his one shot at human flourishing. Against this backdrop, it should seem anything but accidental that the watershed moment in Austerlitz's journey of self-recovery involves counting. This simple conceptual activity is one of the sites at which the fantasy of dehumanized and dispassionate thought is at its most seductive. The fact that Austerlitz begins to reconnect with himself in counting is rightly taken as a sign that he has taken an important step towards claiming his own life.

[25]　Saul Kripke is the most original and influential commentator to defend a reading along these lines. See his now classic *Wittgenstein on Rules and Private Language* (Cambridge, MA: Harvard University Press, 1982). For an in relevant respects similar reading of Wittgenstein by another well-known commentator, see Norman Malcolm, 'Wittgenstein on Language and Rules,' in *Wittgensteinian Themes: Essays 1978–1989* (Ithaca, NY: Cornell University Press, 1995), 145–71.

[26]　See esp. John McDowell's reading of Wittgenstein in 'Non-Cognitivism and Rule-Following' and 'Meaning and Intentionality in Wittgenstein's Later Philosophy,' in *Mind, Value and Reality* (Cambridge, MA: Harvard University Press, 1998), 198–220 and 263–78.

Thus far I have been focusing on passages of Sebald's book that either describe Austerlitz or complement descriptions we are given of him as suffering from a form of emotional stuntedness that amounts to a cognitive limitation. Now I want to turn to considering how this descriptive theme gets addressed in the book's literary strategy. The book employs an intricately interwoven set of literary devices that serve to elicit from readers responses that Austerlitz doesn't at first make and to see things that he only comes to see late in life. Central among the things that come into view for him is that the emotional injury inflicted on him by the rupture with his past is an affliction that keeps him from fully living in a manner that while different from is nevertheless analogous to the way that captive animals of various kinds are prevented from fully living. In the book's fabulous opening passage, the narrator uses rich and expressive language – language that invites responses of a sort the young Austerlitz has difficulty manifesting – to bring out this analogy between Austerlitz and captive animals.

At the book's beginning, the narrator is recounting how he once visited the Nocturama, or nocturnal house, at the Antwerp Zoo and how his visit left him with a clear memory only of one raccoon that, he says, 'sat beside a little stream with a serious expression on its face, washing the same piece of apple over and over again'.[27] The narrator makes it clear not only that he regards the raccoon's repetitive washing as a displaced response to circumstances that keep it from behaving in a more natural way but also that he takes its subjection to these circumstances to represent an injury. He tells us that it seemed to him as if the raccoon 'hoped that all [its] washing, which went far beyond any reasonable thoroughness, would help it to escape the unreal world into which it had arrived, so to speak, through no fault of its own'.[28] After thus evocatively presenting his encounter with the raccoon, the narrator tells us that he walked straight from the Nocturama to the adjoining train station and that, prompted in part by the impression made on him by the fading evening light, he mentally connected the travellers under the station's dome with the inhabitants of the Nocturama. The one traveller who captured his attention, and who thus occupied the same place in his thoughts about this imaginary Nocturama that the raccoon occupied in his thoughts about the real one, was the man – then a stranger to the narrator – whom we soon come to know as Austerlitz.[29] The narrator tells us that Austerlitz was writing busily, taking notes on the hall in which he was seated. We as readers don't yet know the sorts of things about Austerlitz that speak for regarding his immersion in his architectural studies as a strategy for evading life. But the manner in which Austerlitz is thus introduced to us encourages us to connect him with the raccoon and to think of him as resembling it in being constrained to channel his energy into compensatory conduct and in therefore being a fit object of compassion.

Whereas in this first passage the book's narrator evokes for us a perspective from which there is a close analogy between Austerlitz's injured condition and injuries animals can be made to suffer, later in Sebald's book Austerlitz himself repeatedly

---

[27] Sebald, *Austerlitz*, 4.
[28] Ibid.
[29] Ibid., 6–7.

evokes such a perspective. How is Austerlitz, a man harmed in ways that deprive him of self-understanding, capable of doing this? Here it is helpful to bear in mind both how Austerlitz develops during the years Sebald's narrative chronicles and how his way of looking at things is introduced into the narrative. *Austerlitz* is framed as a series of encounters, over roughly 30 years, between Austerlitz and its narrator, and the narrator's account of the encounters possesses a significant temporal complexity even though he presents them in the order in which they occur. The complexity is a function of the sequence in which Austerlitz relates different episodes in his life to the narrator. During the narrator's first meetings with Austerlitz, which take place when Austerlitz is in his early thirties, Austerlitz doesn't mention anything connected with his personal life. The narrator's remaining meetings with Austerlitz take place much later, after Austerlitz has already traced his roots to Prague and begun to understand himself better. But in these further meetings Austerlitz describes his entire life from the time he landed in Wales until his retirement before talking about how he learned about his Czech origins. So, in preserving the temporal order of his interactions with Austerlitz, the narrator gives us a portrait of the 30-something Austerlitz and relays the older Austerlitz's account of himself from age 4½ to nearly 60 without mentioning anything about Austerlitz's earliest years. This means that the sort of information about Austerlitz's past life that would enable us to make more sense of his responses to the world than he can himself early on is withheld from us until quite late in the narrative. It also means that the Austerlitz who is the source of our information about his early years is a man who has in certain respects grown past the disoriented person he once was and is now able to look at and respond to events of his childhood in new ways.

This Austerlitz is a reliable narrative guide, and Sebald's book employs a number of literary techniques to create identification with him and to encourage readers to enter into the perspective he has achieved on his early way of seeing the world. To begin with, although Austerlitz's tale is relayed to us by a narrator, the book is structured in ways that blur the distinction between this narrator and Austerlitz, thereby to a large extent bridging the distance that separates us from Austerlitz. The narrator is a particular, younger man and hence neither omniscient nor disengaged. Like all of the narrators of Sebald's works of prose fiction, he resembles Sebald in a great many respects. Among other things, he is a German expatriate living in England who has significant feelings of dislocation.[30] As a result of his sense of a rupture with his own past, the narrator identifies with Austerlitz's story. He creates a further impression of intimacy with Austerlitz – in a stylistic mannerism arguably inherited from Thomas Bernhard – by moving seamlessly from his own words to Austerlitz's without the kinds of breaks introduced by the use of quotations.[31] Turning now to stylistic peculiarities of Austerlitz's, it is noteworthy that Austerlitz recounts his experiences in long sentences

[30] For Sebald's own remarks on his relation to the Germany of his childhood, see, for example, the interviews in *The Emergence of Memory*, esp. 66–8 and 105–7. See also the partly fictionalized account Sebald gives of his return to the German village in which he spent his earliest years in the long closing section of *Vertigo* (London: Vintage Books, 2002), 169–263.

[31] For Sebald's account of his debt to Bernhard, see, for example, 'A Poem of an Invisible Subject,' an interview with Michael Silverblatt, in *The Emergence of Memory*, 77–86, 82–3.

which, through suggestive associations and other descriptive techniques, evoke his early sense of the world as a bleak and incomprehensible place, a place not only marked but haunted by the spectres of the dead.[32] Complementing these literary techniques are the photographs scattered throughout the book's pages. While some of these photographs are supposed to be images found in books or newspapers or on postcards, most are presented as snapshots taken by the narrator or Austerlitz – snapshots of architectural structures, cityscapes, interior spaces, household objects, artworks and landscapes, as well as of animals and people.[33] Insofar as the snapshots taken by Austerlitz in particular are markers of moments at which the world was present to him, their placement in the narrative represents an occasion to put ourselves in his shoes.[34] Further, insofar as these snapshots present us with glimpses of the past, they contribute to creating the impression of a world that is haunted, as the young Austerlitz's is, by the uncanny presence of days gone by. These are the sorts of things I have in mind in speaking, in reference to Austerlitz's narrative, of different strategies for getting us to enter into and contemplate his early way of experiencing the world.[35]

This brings me back to the specific narrative technique that I mentioned a moment ago, the technique of suggesting a connection between Austerlitz's early condition and the condition of captive animals. There are suggestions along these lines not only in the narrator's own reflections but also in the reflections of Austerlitz's that the narrator relates to us. Consider a passage that comes at a point

[32] For an example of the sort of suggestive association in question, consider the series of echoes, mentioned in note 15 above, among the names 'Austerlitz', 'Auschowitz' and 'Auschwitz'.

[33] We are given the following account of how Austerlitz's photographs fell into the narrator's hands. At one point during the narrative, Austerlitz invites the narrator to his house in London and shows him a collection of black and white photographs that he has taken and collected. Then, later, towards the end of the narrative, when Austerlitz is heading off to learn more about his father and locate the woman he loved and lost, Austerlitz gives the narrator the key to his house and tells him to feel free to study the photographs (*Austerlitz*, 293–4).

[34] Given the use of snapshots in *Austerlitz* (and elsewhere in Sebald's prose narratives) and given the book's many references to the philosophy of Wittgenstein, it is interesting that Wittgenstein once scorned the practice of sharing our personal snapshots with third parties. He spoke of 'those insipid photographs of a piece of scenery which [are] interesting to the person who took it because he was there himself, experienced something, but which a third party looks at with justifiable coldness' (*Culture and Value* (Chicago, IL: University of Chicago Press, 1980), 4e–5e). Despite any initial appearance to the contrary, this remark shouldn't be taken to speak against the way in which snapshots are used in *Austerlitz*. The narrative strategies I've been discussing have the function, within *Austerlitz*, of encouraging identification with Austerlitz and putting us in the position of being not disinterested third parties of the sort Wittgenstein is talking about but rather parties who can enter into Austerlitz's experience and can thus appreciate his snapshots.

[35] Here I need to register a local disagreement with James Wood's in other respects very perceptive remarks on *Austerlitz*. In 'Sent East: Sebald's *Austerlitz*' (*London Review of Books*, 33(19) (6 October 2011), 15–18), Wood represents the structural features of Sebald's narratives that I have been describing – what Wood describes as the 'layering' effected by using a narrator to relate Austerlitz's story – as intended to keep Austerlitz at a distance from us or, in Wood's words, to make Austerlitz 'difficult to get close to' (15). In contrast, I am suggesting that Sebald's complex narrative strategy is designed to make it possible for us to inhabit Austerlitz's mode of experiencing the world and to gain a kind of perspective on it that Austerlitz himself only begins to achieve very late in life. We might say that this narrative strategy makes it possible for us to get closer to Austerlitz than he for most of his life manages to get.

at which Austerlitz has already told the narrator about his journey of discovery to Prague. Austerlitz has now turned to discussing how, after having learned that he and his parents visited the Marienbad spa, he can better understand why the trip that he and Marie de Verneuil took to Marienbad in 1972 was a disaster and why it ultimately led to their separation. After recounting for the narrator how during most of the time he spent with Marie there he was overcome by feelings of despair that didn't make sense to him and that he therefore couldn't illuminate for her, Austerlitz says that on one walk he and Marie took together his attention was caught by a severely decayed dovecote. In it, the bodies of several dead birds lay partially covered in a desiccated mass of droppings and in which the remaining birds survived 'in a kind of senile dementia, coo[ing] at one another in tones of quiet complaint'.[36] Austerlitz represents the condition of the pigeons as intensely wretched. The force of the larger line of thought in which he embeds his account of them is to prompt us to regard his state of emotional confusion as an analogous condition of wretchedness. We are invited to regard the political forces that dislocated him and prevented him from making sense of his life as hateful because they thwart his specifically human development in a manner comparable to the manner in which being confined in cramped and filthy quarters warps the lives of birds.

The ways in which Austerlitz's narrative is designed to get us to appreciate vulnerabilities we humans have as reasoning animals include, in addition to the presentation of analogies between certain harms to animals and certain harms to human beings, the presentation of analogies between the glory of the lives of uninjured animals and the glory of the lives of human beings at fortunate moments at which they enjoy the sort of peace, health and companionship that allows for the unhampered growth of mind. When Austerlitz is discussing occasions on which he felt most at home with himself and most capable of self-understanding, he sometimes gives expression to his sense of the majesty of the moment by describing the splendour of different animals around him. One of these occasions occurs when then school-aged Austerlitz is staying at the house of his best friend Gerald Fitzgerald, the place at which Austerlitz feels most oriented, and he and Gerald have gone on a nighttime excursion with Gerald's naturalist uncle Alphonso. Austerlitz reports that Alphonso entertained the two boys by talking about the natural history of moths and that, after Alphonso lit an incandescent lamp on the high promontory on which they were all sitting, moths began to fly in a sublime display, 'describing thousands of different arcs and spirals and loops, until like snowflakes they formed a silent storm around the light'.[37] On a second occasion, then research student Austerlitz is with Marie, and the two of them are enjoying one of their most tranquil intervals, strolling around the streets of Paris and letting fancy be their guide. Austerlitz's sense of the mysterious grandeur of the outing is brought to his consciousness when he finds himself responding with a surge of overwhelming feeling to a simple band tune

---

[36]   Sebald, *Austerlitz*, 214.
[37]   Ibid., 90.

that he later thinks brings up for him not only hymns from his Welsh childhood but even earlier waltzes and ländler themes. This experience was summed up for him at the time, he says:

> in the image of [a] snow-white goose standing motionless and steadfast among the musicians as long as they played. Neck craning forward slightly, pale eyelids slightly lowered, it listened there in the tent beneath that shimmering firmament of painted stars until the last notes had died away, as if it knew its own future and the fate of its present companions.[38]

The portions of Sebald's book in which Austerlitz expresses a sense of the transient stateliness of uninjured animal life resemble in an important respect the portions, discussed earlier, in which Austerlitz and the narrator evoke the pathos of the plight of captive and damaged animals. These different bits of the book resemble each other in contributing to its efforts to lead readers to see preciousness in some of the very features of human life that leave us exposed to social cataclysm and natural disaster.

This concludes my account of how Sebald's book develops its distinctive, ethically charged and naturalistic image of human beings. The heart of the image is the idea that an emotional handicap of the sort Austerlitz possesses has an essentially cognitive component and represents a real injury. The book tries to get us to acknowledge what is right in this idea, as we have seen, by inviting us to respond in ways that, for most of his life, Austerlitz cannot and by thereby positioning us to recognize that there are aspects of life that are only available from the perspectives our responses afford. It is in large part in this way, by means of a network of literary techniques that engage our sympathies and direct our responses, that Sebald's *Austerlitz* motivates a vision of human beings as animals whose characteristic capacities of mind make us unavoidably vulnerable to political upheavals and natural disasters that uproot us, cutting us off from our pasts.[39]

Let me close with a brief philosophical epilogue. The appeal that *Austerlitz* makes for its distinctive vision of human life does not take the form of an argument, in the

---

[38] Ibid., 275.

[39] While this vision of human fragility gets developed in all of Sebald's prose narratives, *Austerlitz* addresses relevant themes in a particularly powerful manner. A number of generally enthusiastic commentators on Sebald's fiction criticize *Austerlitz* because it focuses on the life-story of one character and because, more than Sebald's other works, it thus contains elements of the traditional novel. The commentators I have in mind claim that in inheriting aspects of traditional novelist form Sebald forces what they see as his bracingly original thought into a shape that cramps it. (See, e.g. the different critical remarks printed in *The Emergence of Memory*, 13, 89, 149 and 169.) Yet these commentators do not offer examples of this supposed cramping, and it is not clear to me that any examples are actually to be found. On the contrary, it seems to me that the different literary techniques of *Austerlitz* I have been discussing – and these include Sebald's more idiosyncratic stylistic techniques as well as more traditional novelistic ones – serve to deepen the book's treatment of themes that run through Sebald's fictional opus. I am inclined to agree with Ruth Klüger when she writes that 'man könnte die Behauptung aufstellen, dass Sebald immer wieder dasselbe Buch geschrieben hat, nur dass es immer besser wurde [one could say that Sebald continually wrote the same book save that it was always getting better]' ('Wanderer Zwischen Falschen Leben,' *Text + Kritik*, Heft 158 (April 2003), 95–102, at 100).

narrow sense in which I have spoken of argument in this essay. Unlike an argument it essentially depends for its effectiveness on eliciting emotional responses. It might accordingly seem right to protest that the book's methods cannot be rationally respectable. Philosophers and critics who thus protest that no non-argumentative mode of instruction can be rationally authoritative tend to reason as follows. They observe that stretches of discourse that direct our responses are not invariably thoughtful and benign and that they are as likely to engage us in ways that encourage the adoption of ignorant and pernicious as well as enlightened outlooks. This observation is then supposed to underwrite the conclusion that, if we are to responsibly allow literature to inform ethical thought, then, prior to engaging with any specific literary work, we need to ask whether the ideas or lines of thought it contains survive the scrutiny of argument.[40] If we applied this method to Sebald's prose narratives, we would evidently be obliged to abandon things I have been saying about their moral interest. To ask whether the thinking traced out in Sebald's works survive the scrutiny of prior argument is to ask whether the connections of thought the works invite us to make are recognizable as justified independently of the sorts of responses that the works seek to cultivate. Given that the strategy that Sebald's narratives – and *Austerlitz* in particular – employ to get us to make new connections of thought depends essentially on techniques for directing our emotional responses and thereby getting us to seriously contemplate a distinctive view of human life, it should be clear that this is not a standard that the narratives are capable of meeting.

Yet there is surely something dubious about the idea that the only way to subject one of Sebald's prose narratives to authoritative assessment is to ask whether the modes of thought with which it presents us receive the sanction of antecedent arguments. This idea is a direct expression of the 'narrower' conception of rationality that, as I brought out earlier in this essay, is entirely foreign to Sebald's thought. The moral psychology that informs Sebald's works of prose fiction, colouring their narrative themes and shaping their literary structures, presupposes the logic of what I have called the 'wider' conception of rationality (viz., the conception on which rational discourse includes more than argument). So there is a straightforward sense in which it is an imposition to insist that the narrower conception of rationality supplies the correct critical standard for measuring the moral interest of Sebald's oeuvre. This doesn't show that Sebald is right to make use of the terms of the wider conception of rationality. But it does show that if we are to have any claim to be taking his fictional enterprise seriously we cannot antecedently insist on scrutinizing it in narrowly rational categories.

Suppose that, drawing on the wider conception of rationality, we do allow that particular literary works may directly inform moral understanding in virtue of their tendencies to engage our feelings in different ways. Suppose, further, that, having made this allowance, we arrive at the conclusion that a specific work thus contributes directly to moral understanding specifically as a literary work. What would it be to show that this conclusion withstands critical examination? We would need to show that the

---

[40] For a clear and representative instance of this line of reasoning, see O'Neill, 'The Power of Example.'

aspects of our lives that the work in question illuminates specifically as a literary work survive reflection on the fact that they only come into view from within the context of the evaluative vision of life it invites us to explore. We would need to step back from the relevant vision and ask whether someone who was unwilling or unable to enter into it would be missing something – or whether, on the contrary, we are dealing with a case of projective illusion. If, for instance, we wanted to address a question about whether we are right to accept the moral vision of Sebald's *Austerlitz*, we would need to start by stepping back from perspectives, of the sorts the book fosters, from which human beings forcibly cut off from their pasts and hindered in their ability to develop emotionally appear to have suffered a particularly grievous injury. We would then need to show that the person who was unwilling or unable to see human life in the pertinent light would genuinely be missing something. To the extent that we thus successfully respond to criticism, we are entitled to regard ourselves as proceeding responsibly in taking from Sebald's *Austerlitz* a striking and morally consequential image of vulnerabilities that we human beings have as the kinds of animals we are.[41]

[41] I am grateful to Cora Diamond, Nathaniel Hupert and Ross Poole for their helpful conversation and correspondence about earlier drafts of this essay. I presented versions of the chapter in the Philosophy Department at the University of Trieste, Italy, at the 2010 inaugural conference of the Nordic Wittgenstein Society at the University of Uppsala, Sweden, and at a 2010 workshop on 'Ethical Forms and Aesthetic Experience' at the Washington Square Hotel in New York. I benefitted from helpful feedback I received on these occasions and would especially like to thank Andrew Benjamin, Jay Bernstein, Piergiorgio Donatelli, Martin Gustafsson, Gregg Horowitz, Katie Kelley, Alison Ross and Marina Sbisá.

Bibliography

Books and articles

Anscombe, G. E. M., *Intention*, 2nd edn (Cambridge, MA: Harvard University Press, 2000).

—, *Human Life, Action and Ethics: Essays*, ed. Mary Geach and Luke Gormally (Exeter: Imprint Academic, 2005).

Araki, Nobuyoshi, *Sentimental Journey, Winter Journey* (Tokyo: Shinchosha, 1991).

—, *Love and Death* (Milan: Silvana Editoriale, 2010).

Aristotle, *De Sensu*, in *De Sensu and De Memoria*, ed. and trans. G. R. T. Ross (Cambridge: Cambridge University Press, 1906), 41–99.

—, *A New Aristotle Reader*, ed. J. L. Ackrill (Oxford: Clarendon Press, 1987).

—, *Politics*, trans. C. D. C. Reeves (Indianapolis: Hackett, 1998).

Arnold, Karen, 'Evaluating Science on Epistemic and Moral Grounds (Formerly, Putting Anthropomorphism in Context' (2001); URL: http://philsci-archive.pitt.edu/373/ (accessed 1 July 2011).

Ascione, Frank R. and Claudia V. Weber, 'Children's Attitudes about the Humane Treatment of Animals and Empathy: One-Year Follow Up of a School-Based Intervention,' *Anthrozoos*, 9 (1996), 188–95.

Balcombe, Jonathan, *Second Nature: The Inner Lives of Animals* (New York: Palgrave Macmillan, 2010).

Beck, Alan and Aaron Katcher, *Between Pets and People: The Importance of Animal Companionship* (West Lafayette, IN: Purdue University Press, 1996).

Bentham, Jeremy, *An Introduction to the Principles of Morals and Legislation* (London: Athlone Press, 1970).

Blair, R. and James R., 'A Cognitive Developmental Approach to Morality: Investigating the Psychopath,' *Cognition*, 57 (1995), 1–29.

—, 'Moral Judgment and Psychopathy,' *Emotion Review*, 3 (2011), 296–8.

Blake, William, *The Complete Poetry and Prose of William Blake*, ed. David V. Erdman (Berkeley, CA: University of California Press, 2008).

Braithwaite, Victoria, *Do Fish Feel Pain?* (Oxford: Oxford University Press, 2010).

Bruner, Jerome, *Child's Talk: Learning to Use Language* (New York: Norton, 1983).

Calarco, Matthew, *Zoographies* (New York: Columbia University Press, 2008).

Cavell, Stanley, *The Claim of Reason: Wittgenstein, Skepticism, Morality, and Tragedy* (Oxford: Oxford University Press, 1979).

—, *This New yet Unapproachable America: Lectures after Emerson after Wittgenstein* (Albuquerque, NM: Living Bath Press, 1989).

—, *Must We Mean What We Say?* (Cambridge: Cambridge University Press, 2002).

—, '"What Becomes of Thinking on Film?" (Stanley Cavell in Conversation with Andrew Klevan),' in *Film as Philosophy: Essays on Cinema after Wittgenstein after Cavell*, ed. Rupert Read and Jerry Goodenough (Basingstoke: Palgrave Macmillan, 2005), 167–209.

—, 'Companionable Thinking,' in Stanley Cavell, Cora Diamond, John McDowell, Ian Hacking and Cary Wolfe, *Philosophy and Animal Life* (New York: Columbia University Press, 2008), 91–126.

Churchill, John, 'If a Lion Could Talk . . .' *Philosophical Investigations*, 12 (1989), 308–24.

Cioffi, Frank, 'Overviews: What Are They of and What Are They For?' in *Seeing Wittgenstein Anew*, ed. William Day and Victor J. Krebs (Cambridge: Cambridge University Press, 2010), 291–313.

Cockburn, David, 'Human Beings and Giant Squids,' *Philosophy*, 69 (1994), 135–50.

Coetzee, J. M., *The Lives of Animals*, ed. Amy Gutmann (Princeton, NJ: Princeton University Press, 1999).

—, *Disgrace* (London: Penguin Books, 2000).

—, *Elizabeth Costello* (New York: Viking, 2003).

Comte, Auguste, *Catéchisme Positiviste* (Paris: Carilian-Goeury and Vor Dalmont, 1852); *The Catechism of Positive Religion*, trans. Richard Congreve (London: Chapman, 1858).

Conrad, Joseph, *The Nigger of the 'Narcissus'* (Garden City, NY: Doubleday, Page & Co., 1897).

Cosmides, Leda and John Tooby, 'Can a General Deontic Logic Capture the Facts of Human Moral Reasoning? How the Mind Interprets Social Exchange Rules and Detects Cheaters,' in *Moral Psychology*, Vol. 1: *The Evolution of Morality: Adaptations and Innateness*, ed. Walter Sinnott-Armstrong (Cambridge, MA: MIT Press, 2008), 53–119.

Crary, Alice, 'Wittgenstein and Ethics: A Discussion With Reference to *On Certainty*,' in *Readings of Wittgenstein's 'On Certainty*,' ed. Danièle Moyal-Sharrock and William H. Brenner (Basingstoke: Palgrave Macmillan, 2005), 275–301.

—, 'Humans, Animals, Right and Wrong,' in *Wittgenstein and the Moral Life: Essays in Honor of Cora Diamond*, ed. Alice Crary (Cambridge, MA: MIT Press, 2007), 381–404.

— (ed.), *Wittgenstein and the Moral Life: Essays in Honor of Cora Diamond* (Cambridge, MA: MIT Press, 2007).

—, 'Literature and Ethics,' in *The International Encyclopedia of Ethics*, ed. Hugh Lafollette (Malden, MA: Wiley-Blackwell, 2012); URL: www.hughlafollette.com/IEE.htm

Crary, Alice and Rupert Read (eds), *The New Wittgenstein* (London: Routledge, 2000).

Cudd, Ann, 'Game Theory and the History of Ideas about Rationality: An Introductory Survey,' *Economics and Philosophy*, 9 (1993), 101–33.

Daly, Beth and Suzanne Suggs, 'Teachers' Experiences with Humane Education and Animals in the Elementary Classroom: Implications for Empathy Development,' *Journal of Moral Education*, 39 (2010), 101–12.

Davidson, Donald, 'On the Very Idea of a Conceptual Scheme,' in *Inquiries into Truth and Interpretation* (New York: Oxford University Press, 1984), 183–98.

—, 'A Coherence Theory of Truth and Knowledge,' in *Truth and Interpretation: Perspectives on the Philosophy of Donald Davidson*, ed. Ernest LePore (Oxford: Blackwell, 1986), 307–19.

de Waal, Frans, 'Morally Evolved, Primate Social Instincts, Human Morality and the Rise and Fall of "Veneer theory",' in *Primates and Philosophers: How Morality Evolved*, ed. Stephen Macedo and Josiah Ober (Princeton, NJ: Princeton University Press, 2006), 1–58.

—, *The Age of Empathy: Nature's Lessons for a Kinder Society* (New York: Harmony, 2009).

de Waal, Frans and Frans Lanting, *Bonobo: The Forgotten Ape* (Berkeley, CA: University of California Press, 1997).

Deigh, John, 'Primitive Emotions,' in *Thinking about Feeling: Contemporary Philosophers on Emotions*, ed. Robert C. Solomon (Oxford: Oxford University Press, 2004), 9–27.

Descartes, René, *Descartes: Philosophical Letters*, ed. and trans. Anthony Kenny (Oxford: Oxford University Press, 1970).

Diamond, Cora, 'Eating Meat and Eating People,' *Philosophy*, 53 (1978), 465–79. Reprinted in *The Realistic Spirit: Wittgenstein, Philosophy, and the Mind* (Cambridge, MA: MIT Press, 1991), 319–34.

—, 'Anything but Argument?' in *The Realistic Spirit: Wittgenstein, Philosophy, and the Mind* (Cambridge, MA: MIT Press, 1991), 291–308.

—, 'The Importance of Being Human,' in *Human Beings*, ed. David Cockburn (Cambridge: Cambridge University Press, 1991), 35–62.

—, *The Realistic Spirit: Wittgenstein, Philosophy, and the Mind* (Cambridge, MA: MIT Press, 1991).

—, 'Ethics, Imagination and the Method of Wittgenstein's *Tractatus*,' in *The New Wittgenstein*, ed. Alice Crary and Rupert Read (London: Routledge, 2000), 149–73.

—, 'Criss-Cross Philosophy,' in *Wittgenstein at Work*, ed. Erich Ammereller and Eugen Fischer (London: Routledge, 2004), 201–20.

—, 'The Difficulty of Reality and the Difficulty of Philosophy,' in Cavell et al., *Philosophy and Animal Life* (New York: Columbia University Press, 2008), 43–89.

Dilman, İlham, *Love and Human Separateness* (Oxford: Blackwell, 1987).

Drury, M. O'C., *The Danger of Words* (London: Routledge and Kegan Paul, 1973).

—, 'Conversations with Wittgenstein,' in *Recollections of Wittgenstein*, ed. Rush Rhees (Oxford: Oxford University Press, 1984), 97–171.

Dupré, John, 'Natural Kinds and Biological Taxa,' *Philosophical Review*, 90 (1981), 66–90.

Ertz, Timo-Peter, *Regel und Witz: Wittgensteinsche Perspektiven auf Mathematik, Sprache und Moral* (Berlin: de Gruyter, 2008).

Eylon, Yuval, 'Virtue and Continence,' *Ethical Theory and Moral Practice*, 12 (2009), 137–51.

Fields, William M., 'Ethnographic Kanzi versus Empirical Kanzi: On the Distinction Between "Home" and "Laboratory" in the Lives of Enculturated Apes,' *Revista di Analisi del Testo Filosofico, Letterario e Figurativeo*, 8 (2007), 171–207.

Fields, William M., Pär Segerdahl and Sue Savage-Rumbaugh, 'The Material Practices of Ape Language Research,' in *The Cambridge Handbook of Sociocultural Psychology*, ed. Jaan Valsiner and Alberto Rosa (Cambridge: Cambridge University Press, 2007), 164–86.

Flack, Jessica C. and Frans de Waal, '"Any Animal Whatever": Darwinian Building Blocks of Morality in Monkeys and Apes,' in *Evolutionary Origins of Morality: Cross Disciplinary Perspectives*, ed. Leonard D. Katz (Thorverton: Imprint Academic, 2000), 1–29.

Frey, Matthias, 'Theorizing Cinema in Sebald and Sebald with Cinema,' in *Searching for Sebald: Photography after W. G. Sebald*, ed. Lise Patt (Los Angeles, CA: Institute of Cultural Inquiry, 2007), 226–41.

Frey, R. G., *Interests and Rights* (Oxford: Oxford University Press, 1980).

Friedman, Michael, 'Exorcising the Philosophical Tradition,' in *Reading McDowell: On 'Mind and World*,' ed. Nicholas H. Smith (London: Routledge, 2002), 25–57.

Gadamer, Hans-Georg, *Truth and Method*, revised trans. Joel Weinsheimer and Donald Marshall (New York: Crossroads, 1992).

Gaita, Raimond, *Romulus, My Father* (Melbourne: Text Publishing, 1998).

—, *The Philosopher's Dog* (Melbourne: Text Publishing, 2002; New York: Random House, 2005).

Gardner, R. Allen and Beatrice T. Gardner, 'Teaching Sign Language to a Chimpanzee', *Science*, 165 (1969), 664–72.

Gauthier, David, *Morals by Agreement* (Oxford: Oxford University Press, 1986).

Gill, Christopher, 'Is There a Concept of Person in Greek Philosophy?' in *Companions to Ancient Thought 2: Psychology*, ed. Stephen Everson (Cambridge: Cambridge University Press, 1991), 166–93.

Glendinning, Simon, *On Being with Others: Heidegger–Derrida–Wittgenstein* (London: Routledge, 1998).

—, *In the Name of Phenomenology* (London: Routledge, 2007).

Goodall, Jane, *In the Shadow of Man* (Boston, MA: Houghton Mifflin, 1988).

Goodall, Jane, R. Brian Sommerville, Leigh E. Nystrom, John M. Darley and Jonathan D. Cohen, 'An fMRI Investigation of Emotional Engagement in Moral Judgment', *Science*, 293 (2001), 2105–8.

Greene, Joshua D. and Jonathan Haidt, 'How (and Where) Does Moral Judgment Work?' *Trends in Cognitive Sciences*, 6 (2002), 517–23.

Griffiths, Paul E., *What Emotions Really Are* (Chicago, IL: Chicago University Press, 1997).

—, 'Emotion on Dover Beach: Feeling and Value in the Philosophy of Robert Solomon', *Emotion Review*, 2 (2010), 22–8.

Hacking, Ian, 'Deflections', in Cavell et al., *Philosophy and Animal Life* (New York: Columbia University Press, 2008), 139–72.

Haidt, Jonathan, 'The Emotional Dog and Its Rational Tail: A Social Intuitionist Approach to Moral Judgment', *Psychological Review*, 108 (2001), 814–34.

Hamilton, William D., 'The Genetical Evolution of Social Behaviour: I and II', *Journal of Theoretical Biology*, 7 (1964), 1–52.

Hamlyn, D. W., *Perception, Learning and the Self: Essays in the Philosophy of Psychology* (Hampshire: Gregg Revivals, 1983).

Hanna, Patricia and Bernard Harrison, *Word and World: Practice and the Foundations of Language* (Cambridge: Cambridge University Press, 2004).

Harding, Elizabeth U., *Kali: The Black Goddess of Dakshineswar* (York Beach, ME: Nicolas-Hays, 1993).

Hare, Brian and Michael Tomasello, 'Chimpanzees Are More Skilful in Competitive Than in Cooperative Cognitive Tasks', *Animal Behaviour*, 68 (2004), 571–81.

Hayes, Keith J. and Catherine Hayes, 'The Intellectual Development of a Home-Raised Chimpanzee', *Proceedings of the American Philosophical Society*, 95 (1951), 105–9.

Hergovich, Andreas, Bardia Monshi, Gabriele Semmler and Verena Zieglmayer, 'The Effects of the Presence of a Dog in the Classroom', *Anthrozoos*, 15 (2002), 37–50.

Hermann, Julia, *Being Moral: Moral Competence and the Limits of Reasonable Doubt*, PhD thesis (Florence: European University Institute, 2011).

—, 'Die Praxis als Quelle moralischer Normativität', in *Moral und Sanktion: Eine Kontroverse über die Autorität moralischer Normen*, ed. Eva Buddeberg and Achim Vesper (Frankfurt: Campus, forthcoming).

Hobbes, Thomas, *Leviathan* (London: Penguin, 1985; Cambridge: Cambridge University Press, 1996; original year of publication 1651).

Hoffman, Martin L., *Empathy and Moral Development: Implications for Caring and Justice* (Cambridge: Cambridge University Press, 2000).

Hughes, Ted, *Ravens* (London: Rainbow Press, 1979).

Hume, David, *Treatise of Human Nature*, ed. L. A. Selby-Bigge (New York: Oxford University Press, 1978).

Hutchinson, Phil, *Shame and Philosophy: An Investigation in the Philosophy of Emotions and Ethics* (Basingstoke: Palgrave Macmillan, 2008).

Hutchinson, Phil, Rupert Read and Wes Sharrock, *There Is No Such Thing as a Social Science: In Defence of Peter Winch* (Aldershot: Ashgate, 2008).

Jaggi, Maya, 'The Last Word,' *The Guardian* (21 December 2001).

Johnston, Paul, *Wittgenstein: Rethinking the Inner* (London: Routledge, 1993).

Josephs, Jeremy, with Susi Bechhöfer, *Rosa's Child: The True Story of One Woman's Search for a Lost Mother and a Vanished Past* (London: I.B. Tauris, 1996).

Joyce, Richard, *The Evolution of Morality* (Cambridge, MA: MIT Press, 2007).

Kant, Immanuel, *Critique of Practical Reason* [1788], trans. Lewis White Beck (Indianapolis: Bobbs-Merrill, 1956).

Kellogg, W. N. and L. A. Kellogg, *The Ape and the Child: A Comparative Study of the Environmental Influence upon Early Behavior* (New York and London: Hafner Publishing Co., 1933).

Klagge, James C., *Wittgenstein in Exile* (Cambridge, MA: MIT Press, 2011).

Klüger, Ruth, 'Wanderer Zwischen Falschen Leben,' *Text + Kritik*, 158 (2003), 95–102.

Kober, Michael, 'On Epistemic and Moral Certainty: A Wittgensteinian Approach,' *International Journal of Philosophical Studies*, 5 (1997), 365–81.

Kotrschal, Kurt and Brita Ortbauer, 'Behavioral Effects of the Presence of a Dog in a Classroom,' *Anthrozoos*, 16 (2003), 147–59.

Kripke, Saul, *Wittgenstein on Rules and Private Language* (Cambridge, MA: Harvard University Press, 1982).

Landy, Joshua, 'A Nation of Madame Bovarys: On the Possibility and Desirability of Moral Improvement Through Fiction,' in *Art and Ethical Criticism*, ed. Garry L. Hagberg (London: Blackwell, 2008), 63–94.

Leahy, Michael P. T., *Against Liberation: Putting Animals in Perspective* (London: Routledge, 1991).

Lelito, Jonathan P. and William D. Brown, 'Complicity or Conflict over Sexual Cannibalism? Male Risk Taking in the Praying Mantis *Tenodera aridifolia sinensis*,' *American Naturalist*, 168 (2006), 263–9.

Lidman, Sara, *Vredens Barn* (Stockholm: Bonniers, 1979).

—, *Naboth's Stone [Nabots sten]*, trans. Joan Tate (Norwich: Norvik Press, 1989).

Linden, Eugene, *The Parrot's Lament and Other True Tales of Animal Intrigue, Intelligence and Ingenuity* (Boston: Dutton, 1999).

Løgstrup, Knud Ejler, *The Ethical Demand* (Notre Dame, IN: University of Notre Dame Press, 1994).

Long, A. A. and D. N. Sedley, *The Hellenistic Philosophers* (Cambridge: Cambridge University Press, 1987).

Malcolm, Norman, *Thought and Knowledge* (Ithaca, NY: Cornell University Press, 1977).

—, 'Wittgenstein on Language and Rules,' in *Wittgensteinian Themes: Essays 1978–1989* (Ithaca, NY: Cornell University Press, 1995), 145–71.

Masson, Jeffrey and Susan McCarthy, *When Elephants Weep: The Emotional Lives of Animals* (New York: Dell, 1995).

Maynard Smith, John, *Evolution and the Theory of Games* (Cambridge: Cambridge University Press, 1982).

McDowell, John, *Mind and World* (Cambridge, MA: Harvard University Press, 1994; paperback edn, 1996).

—, 'Anti-Realism and the Epistemology of Understanding' [1981], in *Meaning, Knowledge, and Reality* (Cambridge, MA: Harvard University Press, 1998), 314–43.

—, *Mind, Knowledge, and Reality* (Cambridge, MA: Harvard University Press, 1998).

—, *Mind, Value and Reality* (Cambridge, MA: Harvard University Press, 1998).

—, 'Response to Crispin Wright,' in *Knowing Our Own Minds*, ed. Crispin Wright, Barry C. Smith and Cynthia Macdonald (Oxford: Clarendon Press, 1998), 47–62.

—, 'Comment on Stanley Cavell's "Companionable Thinking"', in Cavell et al., *Philosophy and Animal Life* (New York: Columbia University Press, 2008), 127–38.

—, 'Avoiding the Myth of the Given,' in *Having the World in View: Essays on Kant, Hegel, and Sellars* (Cambridge MA: Harvard University Press, 2009), 256–72.

—, *The Engaged Intellect: Philosophical Essays* (Cambridge, MA: Harvard University Press, 2009).

—, 'Wittgensteinian "Quietism"', *Common Knowledge*, 15 (2009), 365–72.

McEachrane, Michael, 'Capturing Emotional Thoughts: The Philosophy of Cognitive-Behavioral Therapy,' in *Emotions and Understanding: Wittgensteinian Perspectives*, ed. Ylva Gustafsson, Camilla Kronqvist and Michael McEachrane (Basingstoke: Palgrave, 2009), 81–101.

McKibbin, William F., Todd L. Shackelford, Aaron T. Goetz and Valerie G. Starratt, 'Why Do Men Rape? An Evolutionary Psychological Perspective,' *Review of General Psychology*, 12 (2008), 86–97.

McNicholas, June and Glyn M. Collis, 'Children's Representations of Pets in their Social Networks,' *Child: Care, Health, and Development*, 27 (2001), 279–94.

Melson, Gail F., *Why the Wild Things Are: Animals in the Lives of Children* (Cambridge, MA: Harvard University Press, 2001).

Midgley, Mary, *Beast and Man* (London: Methuen, 1978).

Mill, John Stuart, *Utilitarianism* (London: Longmans, Green, 1882).

Monk, Ray, *Ludwig Wittgenstein: The Duty of Genius* (New York: The Free Press, 1990).

Montgomery, Sy, 'Deep Intellect: Inside the Mind of the Octopus,' *Orion Magazine*, November/December 2011.

Moore, G. E., *Philosophical Papers* (London: Allen and Unwin, 1959).

Moyal-Sharrock, Danièle, *Understanding Wittgenstein's 'On Certainty'* (Basingstoke: Palgrave Macmillan, 2004).

—, 'The Fiction of Paradox: Really Feeling for Anna Karenina,' in *Emotions and Understanding: Wittgensteinian Perspectives*, ed. Ylva Gustafsson, Camilla Kronqvist and Michael McEachrane (Basingstoke: Palgrave, 2009), 165–84.

Muirhead-Thomson, R. C., *Assam Valley: Beliefs and Customs of the Assamese Hindus* (London: Luzac, 1948).

Mulhall, Stephen, 'Ethics in the Light of Wittgenstein,' *Philosophical Papers*, 31 (2002), 293–321.

Murdoch, Iris, *Existentialists and Mystics: Writings in Philosophy and Literature*, ed. Peter J. Conradi (London: Chatto & Windus, 1997; Harmondsworth: Penguin, 1999).

—, 'The Idea of Perfection,' in *Existentialists and Mystics: Writings in Philosophy and Literature*, ed. Peter J. Conradi (London: Chatto & Windus, 1997; Harmondsworth: Penguin, 1999), 299–336.

—, *The Sovereignty of Good Over Other Concepts* (London: Routledge, 2001 [1970]).

Nietzsche, Friedrich, *Twilight of the Idols and The Anti-Christ*, trans. R. J. Hollingdale (London: Penguin, 1968).

—, *On the Genealogy of Morals and Ecce Homo*, trans. R. J. Hollingdale and Walter Kaufmann (New York: Vintage, 1969).

—, *On the Advantage and Disadvantage of History for Life*, trans. Peter Preuss (Indianapolis: Hackett, 1980).

Nussbaum, Martha, *Love's Knowledge: Essays on Philosophy and Literature* (Oxford: Oxford University Press, 1990).

—, *Upheavals of Thought: The Intelligence of Emotions* (Cambridge: Cambridge University Press, 2001).

O'Neill, Onora, 'The Power of Example,' in *Constructions of Reason: Explorations of Kant's Practical Philosophy* (Cambridge: Cambridge University Press, 1989), 165–86.

Ober, Josiah and Stephen Macedo, 'Introduction,' in *Primates and Philosophers: How Morality Evolved*, ed. Stephen Macedo and Josiah Ober (Princeton, NJ: Princeton University Press, 2006), ix–xix.

Orwell, George, *'England, Your England' and Other Essays* (London: Secker & Warburg, 1953).

Paul, Elizabeth S. and James A. Serpell, 'Childhood Pet Keeping and Humane Attitudes in Young Adulthood,' *Animal Welfare*, 2 (1993), 321–37.

Pedersen, Janni and William M. Fields, 'Aspects of Repetition in Bonobo-Human Conversation: Creating Cohesion in a Conversation Between Species,' *Integrative Psychological and Behavioral Science*, 43 (2009), 22–41.

Pedersen, Janni, Pär Segerdahl and William M. Fields, 'Why Apes Point: Indexical Pointing in Spontaneous Conversation of Language-Competent PanHomo Bonobos,' in *Primatology: Theories, Methods and Research*, ed. Emil Potocki and Juliusz Krasiński (Hauppauge, NY: Nova Science Publishers, 2011), 53–71.

Penn, Derek C., Keith J. Holyoak and Daniel J. Povinelli, 'Darwin's Mistake: Explaining the Discontinuity Between Human and Nonhuman Minds,' *Behavioral and Brain Sciences*, 31 (2008), 109–78.

Phillips, D. Z., *Recovering Religious Concepts: Closing Epistemic Divides* (Basingstoke: Macmillan, 2000).

—, 'Wittgensteinianism: Logic, Reality, and God,' in *The Oxford Handbook of Philosophy of Religion*, ed. William J. Wainwright (Oxford: Oxford University Press, 2005), 447–71.

Pippin, Robert, *Henry James and Modern Moral Life* (Cambridge: Cambridge University Press, 2000).

Plato, *Complete Works*, ed. John M. Cooper (Indianapolis: Hackett, 1997).

Pleasants, Nigel, 'Nonsense on Stilts? Wittgenstein, Ethics, and the Lives of Animals,' *Inquiry*, 49 (2006), 314–36.

—, 'Wittgenstein, Ethics and Basic Moral Certainty,' *Inquiry*, 51 (2008), 241–67.

—, 'Wittgenstein and Basic Moral Certainty,' *Philosophia*, 37 (2009), 669–79.

Posner, Richard A., 'Against Ethical Criticism,' *Philosophy and Literature*, 21 (1997), 1–27.

Premack, David, 'Language in Chimpanzee?' *Science*, 172 (1971), 808–22.

Putnam, Hilary, *Words and Life* (Cambridge, MA: Harvard University Press, 1994).

Quine, W. V. O., *Word and Object* (Cambridge, MA: MIT Press, 1960).

—, 'Two Dogmas of Empiricism,' in *From a Logical Point of View* (Cambridge, MA: Harvard University Press, 1961), 20–46.

Raphael, D. D., 'Can Literature be Moral Philosophy?' *New Literary History*, 15 (1983), 1–12.

Regan, Tom, *The Case for Animal Rights* (Berkeley, CA: University of California Press, 1983).

Rhees, Rush, 'The Fundamental Problems of Philosophy' ed. Timothy Tessin, *Philosophical Investigations*, 17 (1994), 573–86.

—, *Wittgenstein and the Possibility of Discourse* (Cambridge: Cambridge University Press, 1998).

—, *Moral Questions*, ed. D. Z. Phillips (Basingstoke: Macmillan, 1999).

Rilke, Rainer Maria, *New Poems*, trans. Edward Snow (San Francisco, CA: North Point Press, 2001).

Robinson, Jenefer, 'Emotion: Biological Fact or Social Construction?' in *Thinking about Feeling: Contemporary Philosophers on Emotions*, ed. Robert C. Solomon (Oxford: Oxford University Press, 2004), 28–43.

Rossano, Matthew J., *Evolutionary Psychology: The Science of Human Behavior and Evolution* (Hoboken, NJ: Wiley, 2003).

Rumbaugh, Duane M., *Language Learning by a Chimpanzee: The LANA Project* (New York: Academic Press, 1977).

Rustin, Michael, 'Innate Morality: A Psychoanalytic Approach to Moral Education,' in *Teaching Right and Wrong: Moral Education in the Balance*, ed. Richard Smith and Paul Standish (Staffordshire: Trentham Books, 1997), 75–91.

Sanchez, Lisa E., 'How Homo Academicus Got His Name and Other Just-So Stories,' *Gender Issues*, 18 (2000), 83–103.

Savage-Rumbaugh, E. Sue, *Ape Language* (New York: Columbia University Press, 1986).

Savage-Rumbaugh, E. Sue and Roger Lewin, *Kanzi: The Ape at the Brink of the Human Mind* (New York: Wiley, 1994).

Savage-Rumbaugh, E. Sue, Stuart G. Shanker and Talbot J. Taylor, *Apes, Language and the Human Mind* (Oxford: Oxford University Press, 1998).

Savage-Rumbaugh, E. Sue, Jeannine Murphy, Rose A. Sevcik, Karen E. Brakke, Shelley L. Williams and Duane M. Rumbaugh, *Language Comprehension in Ape and Child*, Monographs of the Society for Research in Child Development, 58(3–4) (Oxford: Blackwell, 1993).

Scanlon, Thomas M., *What We Owe to Each Other* (Cambridge, MA: Harvard University Press, 1998).

Schwart, Lynne Sharon, *The Emergence of Memory: Conversations with W. G. Sebald* (New York: Seven Stories Press, 2007).

Sebald, W. G., *The Emigrants*, trans. Michael Hulse (New York: New Directions, 1997).

—, *Rings of Saturn*, trans. Michael Hulse (London: Vintage, 1998).

—, *Austerlitz*, trans. Anthea Bell (New York: Modern Library, 2001).

—, *After Nature*, trans. Michael Hamburger (New York: Random House, 2002).

—, *Vertigo*, trans. Michael Hulse (London: Vintage Books, 2002).

—, *On the Natural History of Destruction*, trans. Anthea Bell (New York: Modern Library, 2003).

Segerdahl, Pär, William M. Fields and Sue Savage-Rumbaugh, *Kanzi's Primal Language: The Cultural Initiation of Primates into Language* (Basingstoke: Palgrave Macmillan, 2005).

Sellars, Wilfrid, 'Empiricism and the Philosophy of Mind,' in *Minnesota Studies in the Philosophy of Science*, Vol. 1, ed. Herbert Feigl and Michael Scriven (Minneapolis: University of Minnesota Press, 1956), 253–329.

Sharpe, Lynne, *Creatures Like Us?* (Exeter: Imprint Academic, 2005).

Sherman, Gary D. and Jonathan Haidt, 'Cuteness and Disgust: The Humanizing and Dehumanizing Effects of Emotion,' *Emotion Review*, 3 (2011), 245–51.

Shostak, Marjorie, *Nisa: The Life and Words of a !Kung Woman* (Cambridge, MA: Harvard University Press, 1981).

Singer, Peter, *Animal Liberation: A New Ethics for our Treatment of Animals* (New York: Random House, 1975; Harper Perennial, 2009).

—, *Practical Ethics* (Cambridge: Cambridge University Press, 1993).

—, 'Reflections,' in J. M. Coetzee, *The Lives of Animals*, ed. Amy Gutmann (Princeton, NJ: Princeton University Press, 1999), 85–92.

—, *The Expanding Circle: Ethics, Evolution, and Moral Progress* (Princeton, NJ: Princeton University Press, 2011 [1981]).

Slote, Michael, *Moral Sentimentalism* (Oxford: Oxford University Press, 2010).

Smith, Adam, *The Theory of Moral Sentiments*, ed. Knud Haakonsen (Cambridge: Cambridge University Press, 2002).

Solomon, Robert C., 'Back to Basics: On the Very Idea of "Basic Emotions",' *Journal for the Theory of Social Behaviour*, 32 (2002), 115–44.

Sorabji, Richard, *Animal Minds and Human Morals* (London: Duckworth, 1993).

Stemmer, Peter, *Handeln zugunsten anderer: Eine moralphilosophische Untersuchung* (Berlin: de Gruyter, 2000).

Steutel, Jan and Ben Spiecker, 'Cultivating Sentimental Dispositions through Aristotelian Habituation,' *Journal of Philosophy of Education*, 38 (2004), 531–49.

Stone, Valerie, 'The Moral Dimensions of Human Social Intelligence: Domain-Specific and Domain-General Mechanisms,' *Philosophical Explorations*, 9 (2006), 55–68.

Stroll, Avrum, *Moore and Wittgenstein on Certainty* (Oxford: Oxford University Press, 1994).

Taglialatela, Jared P., Sue Savage-Rumbaugh and Lauren A. Baker, 'Vocal Production by a Language-Competent *Pan paniscus*,' *International Journal of Primatology*, 24 (2003), 1–17.

Taylor, Charles, *Human Agency and Language* (Cambridge: Cambridge University Press, 1985).

Terrace, Herbert S., *Nim* (New York: Knopf, 1979).

Terrace, Herbert S., L. A. Petitto, R. J. Sanders and T. G. Bever, 'Can an Ape Create a Sentence?' *Science*, 206 (1979), 891–902.

Thoreau, Henry D., *Walden: An Annotated Edition*, ed. Walter Harding (New York: Houghton Mifflin, 1995).

Thornhill, Randy, 'Sexual Selection and Paternal Investment in Insects,' *American Naturalist*, 110 (1976), 153–63.

—, 'Rape in Panorpa Scorpionflies and a General Rape Hypothesis,' *Animal Behavior*, 28 (1980), 52–9.

Thornhill, Randy and John Alcock, *The Evolution of Insect Mating Systems* (Cambridge, MA: Harvard University Press, 1983).

Thornhill, Randy and Craig T. Palmer, *A Natural History of Rape: Biological Bases of Sexual Coercion* (Cambridge, MA: MIT Press, 2000).

Tomasello, Michael, *The Cultural Origins of Human Cognition* (Cambridge, MA: Harvard University Press, 1999).

—, 'Why Don't Apes Point?' in *Roots of Human Sociality: Culture, Cognition and Interaction*, ed. Nicholas J. Enfield and Stephen C. Levinson (London: Berg, 2006), 506–24.

Tooby, J., L. Cosmides, A. Sell, D. Liebermann and D. Sznycer, 'Internal Regulatory Variables and the Design of Human Motivation: A Computational and Evolutionary Approach,' in *Handbook of Approach and Avoidance Motivation*, ed. Andrew J. Elliot (Mahwah, NJ: Erlbaum, 2008), 251–71.

Toth, Nicholas, Kathy Schick and Sileshi Semaw, 'A Comparative Study of the Stone Tool-Making Skills of *Pan, Australopithecus, and Homo sapiens*,' in *The Oldowan: Case Studies into the Earliest Stone Age*, ed. Nicholas Toth and Kathy Schick (Bloomington, IN: CRAFT Press, 2003), 155–222.

Toth, Nicholas, Kathy D. Schick, E. Sue Savage-Rumbaugh, Rose A. Sevcik and Duane M. Rumbaugh, 'Pan the Tool-Maker: Investigations into Stone Tool-Making and Tool-Using Capabilities of a Bonobo (*Pan paniscus*),' *Journal of Archeological Science*, 20 (1993), 81–91.

Trivers, Robert L., 'Parental Investment and Sexual Selection,' in *Sexual Selection and the Descent of Man: 1871–1971*, ed. Bernard Campbell (Chicago, IL: Aldine, 1972), 136–79.

von Grundherr, Michael, *Moral aus Interesse: Metaethik der Vertragstheorie* (Berlin: de Gruyter, 2007).

Wallman, Joel, *Aping Language* (Cambridge: Cambridge University Press, 1992).

Ward, Tony and Richard Siegert, 'Rape and Evolutionary Psychology: A Critique of Thornhill and Palmer's Theory,' *Aggression and Violent Behaviour*, 7 (2002), 145–68.

Weil, Simone, *Simone Weil: An Anthology*, ed. Siân Miles (New York: Weidenfeld and Nicholson, 1986).

Westermarck, Edward, *The Origin and Development of the Moral Ideas*, 2nd edn, 2 vols (London: Macmillan, 1912 and 1917).

—, *Ethical Relativity* (London: Kegan Paul, Trench, Trubner & Co., 1932).

Wierzbicka, Anna, 'Everyday Conceptions of Emotion: A Semantic Perspective,' in *Everyday Conceptions of Emotion: An Introduction to the Psychology, Anthropology and Linguistics of Emotion*, ed. James A. Russell, José-Miguel Fernández-Dols, Antony S. R. Manstead and J. C. Wellenkamp (Dordrecht: Kluwer, 1995), 17–47.

Wilde, Oscar, *Collected Works of Oscar Wilde* (Ware: Wordsworth Editions, 1997).

Williams, Bernard, *Ethics and the Limits of Philosophy* (London: Fontana, 1985).

Williams, Michael, 'Wittgenstein, Truth and Certainty,' in *Wittgenstein's Lasting Significance*, ed. Max Kölbel and Bernhard Weiss (London: Routledge, 2004), 247–81.

Wilson, Catherine, 'Disgrace: Bernard Williams and J. M. Coetzee,' in *Art and Ethical Criticism*, ed. Garry Hagberg (Oxford: Blackwell, 2008), 144–62.

Winch, Peter, *The Idea of a Social Science and Its Relation to Philosophy* (London: Routledge & Kegan Paul, 1958; 2nd edn, 1990).

—, 'Understanding a Primitive Society,' in *Ethics and Action* (London: Routledge, 1972), 8–49.

Wittgenstein, Ludwig, *Philosophical Investigations*, trans. G. E. M. Anscombe (Oxford: Blackwell, 1953; 2nd edn, 1958; 3rd edn, 1967).

—, 'A Lecture on Ethics,' *Philosophical Review*, 74 (1965), 1–12. Reprinted in *Ludwig Wittgenstein: Philosophical Occasions, 1912–1951*, ed. James C. Klagge and Alfred Nordmann (Indianapolis: Hackett, 1993), 115–55.

—, *The Blue and Brown Books: Preliminary Studies for the 'Philosophical Investigations'* (New York: Harper Torchbooks, 1965; 2nd edn, Malden, MA: Blackwell, 1969).

—, *Lectures and Conversations on Aesthetics, Psychology and Religious Belief*, ed. Cyril Barrett (Oxford: Blackwell, 1966).

—, *Philosophical Grammar*, ed. Rush Rhees, trans. Anthony Kenny (Berkeley, CA: University of California Press, 1974).

—, *Tractatus Logico-Philosophicus*, trans. D. F. Pears and B. F. McGuinness, rev. edn (London: Routledge, 1974).

—, *Remarks on the Foundations of Mathematics*, ed. G. H. von Wright, Rush Rhees and G. E. M. Anscombe, trans. G. E. M. Anscombe, 3rd edn (Oxford: Blackwell, 1978).

—, *Remarks on Frazer's 'Golden Bough,'* trans. A. C. Miles, revised by Rush Rhees (Bishopstone: Brynmill Press, 1979).

—, *Notebooks, 1914–1916*, ed. G. H. von Wright and G. E. M. Anscombe, trans. G. E. M. Anscombe, 2nd edn (Oxford: Blackwell, 1979).

—, *On Certainty*, ed. G. E. M. Anscombe and G. H. von Wright, trans. Denis Paul and G. E. M. Anscombe (New York: Harper Torchbooks, 1972; Oxford: Blackwell, 1979).

—, *Culture and Value*, ed. G. H. von Wright and Heikki Nyman, trans. Peter Winch (Chicago, IL: University of Chicago Press, 1980).

—, *Remarks on the Philosophy of Psychology*, 2 vols (Chicago, IL: University of Chicago Press, 1980).

—, *Zettel*, trans. G. E. M. Anscombe (Berkeley, CA: University of California Press, 1967; 2nd edn, Oxford: Blackwell, 1981).

—, *Last Writings on the Philosophy of Psychology*, Vol. I (Chicago, IL: University of Chicago Press, 1982).

—, *Last Writings on the Philosophy of Psychology*, Vol. II (Oxford: Blackwell, 1992).

—, 'Remarks on Frazer's *Golden Bough,*' trans. John Beversluis. In *Philosophical Occasions, 1912–1951*, ed. James C. Klagge and Alfred Nordmann (Indianapolis: Hackett, 1993), 115–55.

—, *The Big Typescript: TS 213*, ed. and trans. C. Grant Luckhardt and Maximillian A. E. Haue (Oxford: Blackwell, 2005).

—, *Philosophical Investigations*, 4th edn, trans. G. E. M. Anscombe, P. M. S. Hacker and Joachim Schulte (Malden, MA: Wiley-Blackwell, 2009).

—, 'Philosophy of Psychology – A Fragment,' in *Philosophical Investigations*, 4th edn, trans. G. E. M. Anscombe, P. M. S. Hacker and Joachim Schulte (Malden, MA: Wiley-Blackwell, 2009), 182–243.

Wood, James, 'Sent East: Sebald's *Austerlitz,*' *London Review of Books*, 33(19) (6 October 2011), 15–18.

Young, Liane, Joan Albert Camprodon, Marc Hauser, Alvaro Pascual-Leone and Rebecca Saxe, 'Disruption of the Right Temporoparietal Junction with Transcranial Magnetic Stimulation Reduces the Role of Beliefs in Moral Judgments,' *PNAS (Proceedings of the National Academy of Sciences of the United States of America)*, 107 (2010), 6753–8.

Young, Liane, Antoine Bechara, Daniel Tranel, Hanna Damasio, Marc Hauser and Antonio Damasio, 'Damage to Ventromedial Prefrontal Cortex Impairs Judgment of Harmful Intent,' *Neuron*, 65 (2010), 845–51.

Films

George, Sally (director), *Whatever Happened to Susi?* (BBC Television, 1991).

Jeunet, Jean-Pierre (director), *Alien: Resurrection* (Twentieth Century Fox, 1997).

Niio, Genya (director), *Kanzi I: Kanzi, an Ape of Genius* (Tokyo: NHK, 1993).

— (director), *Kanzi II* (Tokyo: NHK, 2000).

Resnais, Alain (director), *L'année Dernière à Marienbad [Last Year at Marienbad]* (France: Cocinor, 1961).

Savage-Rumbaugh, E. Sue, *Bonobo People: The Story of Research into Language and Cognition of a Family of Bonobos at the Language Research Center* (Atlanta, GA: Language Research Center, NHK, Public Sphere, 1994).

Index

www.ingramcontent.com/pod-product-compliance
Lightning Source LLC
Chambersburg PA
CBHW071855270326
41929CB00013B/2243